Crystallization Modalities in Polymer Melt Processing

Hermann Janeschitz-Kriegl

Crystallization Modalities in Polymer Melt Processing

Second Edition

 Springer

Hermann Janeschitz-Kriegl
Johannes Kepler University
Linz
Austria

ISBN 978-3-030-08426-4   ISBN 978-3-319-77317-9   (eBook)
https://doi.org/10.1007/978-3-319-77317-9

1st edition: © Springer-Verlag/Wien 2010
2nd edition: © Springer International Publishing AG 2018
Softcover re-print of the Hardcover 2nd edition 2018

Printed on acid-free paper

This Springer imprint is published by Springer Nature
The registered company is Springer International Publishing AG
The registered company address is: Gewerbestrasse 11, 6330 Cham, Switzerland

Preface to the Second Edition

It seems that the date, when a manuscript is ready for print, is not always the moment, when the author is convinced that no further improvements will become necessary. This experience overcame the author in 2008, when the manuscript of the first edition was sent to the publishing house. However, as no author can be sure of the success of the first edition, he can also not be sure that the publisher will ever encourage a second edition. As a consequence, the present author decided to publish several additional papers in refereed journals in the hope that people would find them. Fortunately, Springer decided meanwhile to ask the author in 2014 for a second edition. A desirable solution of the problem seemed imminent. However, because of the high age of the author he tried to find support for the preparation of a second edition (an electronic version). Unfortunately, misunderstandings caused a considerable delay. Finally, the daughter of the author (a physicist) came to his help. He also should mention his long-standing cooperator Dr. Ewa Ratajski for her renewed assistance in the required experiments and the Austrian Science Foundation for further financial support.

In fact, we were happy in developing a new technique. As is well-known, shear flow is the prevailing type of flow in many processing techniques for the manufacturing of plastic articles. However, in general only short-term shearing is applied. But the achievable total rectilinear shear flow between glass or metal strips is of interest in widening the experimental window. Such a flow can happen by the application of a pressure gradient in flow direction or by the relative movement of one wall with respect to the other wall in creating a homogeneous shear rate. But the lengths of those strips can in practice not be much larger than ten centimeters. And with shear flow one has to stop the movement of the moveable strip after about nine centimeters, leaving a residual length for the fluid of not much more than one centimeter. As a reasonable sample thickness of one millimeter is required, a total shear of at best ninety can be reached. On the other hand, one can achieve a rather large number of revolutions in any of the usual rotational viscometers. However, as a consequence of the curvature of the flow lines one gets radial secondary flow in these machines. With the usual small radii of up to two centimeters of those machines, secondary flow becomes of influence already after a few revolutions. This fact becomes particularly disturbing, if crystallization is initiated by the flow.

But recently we were able to develop a parallel plate rheometer of a much larger effective radius in reducing the curvature of the flow lines considerably. The pertinent polymer sample is ring-shaped and can be cut from a pre-molded sheet. In our machine, the outer diameter of this sample is 18 cm and the inner diameter is 14 cm. Apparently, with this sample the curvature of the flow lines is sufficiently reduced. In fact, one can get with this machine practically undisturbed total shears up to more than 3000. Interestingly enough one gets highly oriented structures also at low shear rates, if the shearing is extended. The birefringence of the extracted sample, which is obtained after a quench, increases linearly with the applied shear. No saturation has been found so far.

Linz, Austria   Hermann Janeschitz-Kriegl
November 2017

Preface to the First Edition

Structure formation in crystallizing polymers, as occurring during processing, has not been treated so far in a coherent form. This fact explains why this monograph is written as the first book devoted to this subject. A quarter of a century ago, the underdevelopment of this subject was obvious. Trial and error dominated. In fact, other apposite subjects as polymer melt rheology or heat transfer had reached high levels. A great number of books have been devoted to them. Mold filling of amorphous polymers and the solidification of these polymers by vitrification can nowadays be simulated numerically with a high degree of accuracy. In the solidified samples even residual stresses and corresponding birefringence effects can accurately be calculated.[1]

However, semicrystalline polymers, which form the majority of industrial polymers, have been excluded from these considerations for good reasons. In fact, great uncertainties existed about the formation of quality determining crystalline structures. In particular, polyolefins suffered from this shortcoming. In 1983, this fact instigated the polymer research group at the Johannes Kepler University in Linz to start with pertinent activities. The urgency of this kind of studies becomes evident, if advantages and hitches of these polymers are considered.

1. Versatility of processing: Injection molding into a great variety of shapes and sizes, from thin walled bakers to garden chairs, not to forget pipe and profile extrusion, cable coating, fiber spinning, film blowing.
2. Product qualities: Ductility, low density, good electric insulation, corrosion resistance, surface quality.
3. Processing parameters: No molding process is thinkable without pressurization, flow, and quenching. But fast flow has an unavoidable influence on the anisotropy of the products. The grain size is influenced by deformation rate, pressure, and speed of cooling.

---

[1]Baaijens FTP (1991) Calculation of residual stresses in injection molded products. Rheol Acta 30:284–299.

4. Effects on end-use properties: Unfortunately, so far there are mostly only qualitative insights into the consequences of undesired structures. One knows of the anisotropy of shrinkage, warping, insufficient dimensional stability, oriented craze, and crack development (a side effect is split fiber formation) anisotropy of strength and elasticity.
5. Remedies: First of all, the origins of all kinds of structure elements must be explored. Of course, a solid theoretical foundation for the connection between those crystalline structures and end-use properties would be desirable. For the moment, however, this book is devoted to an elucidation of the background. The reader must not be disappointed, if only a partial goal is envisaged. Instead, he should be content with the fascinations, which nature provides even in such a prosaic subject as polymer processing.

In fact, some of the fundamental processes had not even been recognized. Striking examples are the enormous increase of the number density of stable nuclei (by many decades) as a result of fast cooling, of fast flow, or of pressurization and the role of local alignments in this matter (how to explain self-seeding!). Also, nobody has ever expected great differences in the stability of nuclei of varying origin, or the absence of sporadic nucleation in quiescent melts of flexible polymers. The surprisingly slow relaxation of these nuclei had not been observed previously. Other authors have later adopted this technique for shearing experiments under the name "shear pulse." Some of these authors found this technique so evident that they forgot to quote us. In an elegant way, this technique separates nucleation during flow and subsequent growth.

In fact, only strategic experimentation could lead to those remarkable results. This meant that from the beginning our creativeness was directed on the construction of equipment, which could not be on the market. In fact, experimental conditions had to be chosen close enough to those prevailing in practical processing. Otherwise, the obtained results could not be relevant enough. For the evaluation of the results, however, we could easily live with conventional techniques as birefringence, optical and electron microscopy, and, occasionally, IR and X-ray techniques. In addition, our Viennese friends started with a mathematical technique for the correct description of the process of crystallization.[2] It goes without saying that this process occurs on the surfaces of already formed crystalline areas, a fact which has not been taken into account so far.

In some respect, this monograph has the character of a detective story. This means that this book should not be considered as an ordinary review. In the first instance, it should be seen as a means to promote a new train of thoughts. As a consequence, papers from supportive authors have preferentially been quoted and discussed. The author wants to thank Prof. Dealy (Montreal) for drawing attention to a serious flaw in the manuscript. It is also written in memoriam of D. W. van Krevelen and P. J. Flory for their early encouragement.

---

[2] Schneider W, Köppl A, Berger J (1988) Non-isothermal crystallization of polymers. Int Polym Proc 2:151–154.

It goes without saying that numerical simulations, as far as crystallization is concerned, can only be questionable at the moment. Admittedly, there are some hopeful exceptions. In fact, the similarity of the shear impulse technique with the injection molding process cannot be overlooked, even if the latter technique is not isothermal. So one can hope that old experiences with mold filling can be of some help in future interpretations. (Can flow during mold filling be considered as nearly undisturbed by incipient crystallization?)

Finally, the hope is uttered that this monograph will find a sufficient number of benevolent readers. It should serve as a justification for all the energy put into the subject by a small group of people at Linz University during the last 25 years, where work orders of the industry brought satisfaction but also considerable delay.

Linz, Austria   Hermann Janeschitz-Kriegl
August 2009

Contents

About the Author

Hermann Janeschitz-Kriegl professor emeritus from Johannes Kepler University Linz, was born in Graz, Austria, October 22, 1924. He studied chemistry at Graz University where he graduated with a doctoral thesis on cellulose in 1951. From 1952–1968 he worked at TNO Delft, the Netherlands, on several polymer science and engineering subjects. From 1968–1978 he was a full professor at Delft University. He became a Fellow of the Plastics Institute London, UK, in 1968. Since 1978 he has been professor of physical chemistry at Linz University. He is the author of the monograph "Polymer Melt Rheology and Flow Birefringence", Springer 1983, and the first edition of the present book, Springer 2009. In 1994 he received the Honorary Doctorate from Leoben Unversity, and in 2003 the Honorary Membership of the German Society of Rheology.

Chapter 1
Required Basic Achievements

1.1 Interaction of Three Transport Phenomena: Heat Transfer, Flow and Crystallization Kinetics

It is evident that the three transport phenomena, which are mentioned in the title of this section, are involved in structure formation during processing. In particular, flow is always engaged in mold filling. In one respect flow causes macroscopic heat and momentum transport. But it has also an enormous influence on the crystallization kinetics. Actually, crystallization is the consequence of transport on a micro-scale. It is rendered possible by rearrangements of molecules. However, these rearrangements are favored by flow. This seems obvious. But nobody would have expected that the influence of flow should be so tremendous. In fact, the biggest surprise for us was that the number density of (apparently athermal) nuclei could be enhanced by many tens, if shear or extensional flows were applied for short time spans to melts of industrial polypropylenes[1–3], which were undercooled to temperatures below the melting temperature of their spherulites (see Fig. 1.3). A similar big effect could be attained by rapid quenches of quiescent melts to a series of much lower temperatures. The details of these experiments will be described later in Sect. 2.1.1. However, a demonstration of the overwhelming effects seems important. For the purpose, Fig. 1.1 is introduced.

This figure is a three dimensional plot in perspective. The left horizontal axis gives the temperatures, to which the samples have been rapidly quenched from a temperature well above the equilibrium melting point. The right horizontal axis gives the specific mechanical works applied to the samples during shear and extensional flows at some not too low temperatures. On the vertical axis the obtained number densities of nuclei are plotted. Because of the enormous influences of temperature or mechanical work, a logarithmic scale had to be chosen. With respect to the temperature axis one should not forget that the equilibrium melting point of PP is at 212 °C [4]. But this means that the temperature axis must be doubled in length to get this point on the line. One could guess, how low the

© Springer International Publishing AG 2018
H. Janeschitz-Kriegl, *Crystallization Modalities in Polymer Melt Processing*,
https://doi.org/10.1007/978-3-319-77317-9_1

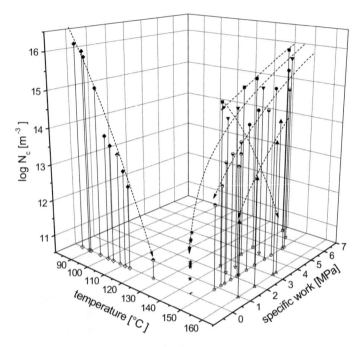

Fig. 1.1 A three dimensional plot of the (logarithm) of the number density of nuclei against the temperatures of fast quenches (left horizontal axis) and against the applied specific mechanical works (right horizontal axis) for an industrial PP according to [2, 3]. Courtesy of Springer Verlag

number density of these type of nuclei would be near this temperature. But also the work axis should be extended to 25 MPa for including the transition to thread-like nuclei. These remarks are made in promoting a feeling for the enormous range of relevant variables. The nature of the nuclei (sporadic, thermal or athermal) will be discussed later. The same holds for the use of specific work instead of flow time and external stress. Here one can suffice by noticing that the rate of specific work is the product of the external stress, which causes the orientation of the molecules (see the stress optical rule [5]), and the rate of deformation, which is proportional to the probability of an encounter of oriented molecules [6].

Figure 1.1 teaches us that there must be a tremendous interaction between the said variables. In particular, the crystallization kinetics, as expressed in Fig. 1.1 by the appearance of nuclei, is strongly dependent on flow and temperature. The second parameter of interest is the growth speed of crystalline domains. Unfortunately, it has not been possible so far to develop a method for a determination of the influence of flow on the growth speed. For quiescent melts of a series of polypropylenes the growth speed of spherulites is plotted as a function of temperature in Fig. 1.2 [7]. A description of the pertinent experiments is given in

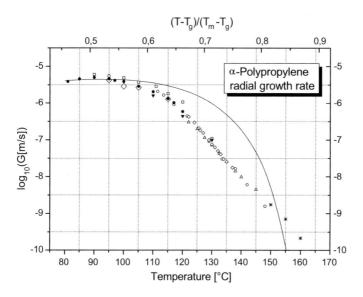

Fig. 1.2 The growth speed of spherulites, as observed in quiescent melts, is given as a function of the temperature of crystallization for a series of polypropylenes according to various sources given in Ref. [7]. The low temperature values were exclusively obtained in our laboratory [7]. Courtesy of Springer Verlag

Sect. 2.1.2. Many external sources, however, have been used for the upper range of temperatures.

As in many slower crystallizing polymers, a maximum is observed half-way down to the glass transition temperature[8]. The full line refers to a theory by Van Krevelen, which is grafted on results of Ref. [8] (see Sect. 2.1.2). However, in this connection it must be admitted that for flow, which is not interrupted, the speed of lateral growth on lengthy crystalline entities (the so-called shish-kebabs) has not yet received a proper treatment. The reason for this omission lies in large experimental difficulties. In fact, spherulites do not grow during continued flow. In contrast, those spherical entities can grow slowly after cessation of short term flow. The pertinent number densities of spherulites correspond with the number densities of nuclei created by this short term flow. Apparently the mutual distances between the said nuclei was still large, when flow was stopped. See the right side of Fig. 1.1.

At this point it seems important, however, to discern between the effects of nucleation and growth on the process and its consequences. On one side the number and shape of nuclei has a large influence on the morphology, which can be observed in the cross-sections of the final product. On the other side, growth speeds have a large influence on the course of the overall crystallization. In fact, as will be shown below, only the first power of the number density of nuclei shows up in the relevant equations, whereas a higher power, mostly the third power, characterizes the influence of the growth speed. One should not overlook the importance of an inherent proneness of some polymers (see HDPE) to fast crystallization. In fact, a fast evolution of latent heat can be quite obstructive to the cooling process and will level up inside the sample the crystallization temperature, which otherwise is lowest near the cooled surface and increases with the distance from this surface. However, the pertinent expectations must be quelled at this stage. To some extent, all these problems will pass the review in later sections.

At the end of this section the T, S-diagram of polypropylene is given in Fig. 1.3 (T … absolute temperature, S … specific entropy, see Ref. [9]).

The advantage of this uncommon diagram is that one can directly read the relevant temperatures from the graph. The entropy as the independent variable can easily be calculated for a great number of polymers from well-known thermodynamic tables (see e.g. [10]). The figure describes two types of transitions from the fluid to the solid. On the right side one has the common liquidus line "l". On the left side one has two solidus lines "s", dependent on the type of the transition. The level of the upper horizontal line is that of the equilibrium melting point of ideal crystals [4], containing macromolecules in their stretched conformation. As is well-known there is no way to realize this transition directly. For this purpose one would have to start with sporadic nuclei, in which the molecules were already stretched. But the probability for those nuclei is practically zero in view of the coiled conformations of the melt. Ideal crystals can only be obtained in a slow re-crystallization process, preferably under high pressure and at temperatures close to the equilibrium melting point [11, 12].

The lower level of transition, as indicated in Fig. 1.3 by two parallel horizontal lines, corresponds to the melting or formation of spherulites. These spherulites contain lamellae of finite thickness. In such a lamella the macromolecules are arranged in a direction perpendicular to the extension of the lamella [13]. A double horizontal line is drawn because the melting temperature of the spherulites is not so well defined. The origin of these lamellae will be discussed in more detail in Sect. 2.2.3. In this section the role of local alignments as pre-stages of athermal nuclei will be brought in. With a not too low average molar mass the length of any

Fig. 1.3 The T,S-diagram of PP. For details see the text [9]. Courtesy of Springer, Vienna

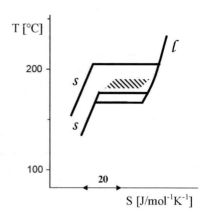

molecule will surpass the length of any alignment by far. As a consequence, macromolecules will protrude from the body of the lamella on both sides or will be bent back into the lamella [14]. Also in this case no sporadic nucleation can be expected.

Looking at Fig. 1.3 one recognizes two temperature ranges, one between the two transitions and the other below the lower transition. It will turn out that in the upper range of temperatures sporadic nucleation can indeed be initiated by the action of flow, apparently because of the uncoiling of the molecules. In the range of temperatures below the melting range of spherulites (below the double line) athermal nuclei dominate and cause very different mechanisms for crystallization and, in particular, for flow induced crystallization. In this connection the stability of flow induced thread-like nuclei is remarkable. In fact, if those thread-like nuclei are formed in fast flows at temperatures close to the equilibrium melting point [15], they show a pronounced relaxation phenomenon. However, the length of the relaxation times increases so fast with decreasing temperature that close to the melting temperature of the spherulites these thread-like nuclei can be considered as practically stable, certainly with respect to the time scale of polymer processing. The temperature range of those stabilized nuclei is indicated by the shaded area in Fig. 1.3 (see also Ref. [16]). A similar kind of stability holds for nuclei, which are formed in the process of self-nucleation, as occurring in quiescent situations at temperatures around the melting temperature of spherulites [17]. These facts permit us to believe in the predominance of athermal nuclei in the temperature range below the melting temperature of spherulites, i.e. below the horizontal double line in Fig. 1.3.

Only years after the completion of the manuscript of the first edition it occurred to the author that the interpretation of the above Fig. 1.1 must necessarily be supplemented. So far the attention was concentrated on the enormous influence of flow on the number density of nuclei. In fact, if one tries to extrapolate the line for zero specific work (for the quiet melt)to higher temperatures, one finds for 145 °C a density of less that 10^9 m^{-3}. In this estimate impurities are disregarded. If the extrapolated density is compared with the density, which is obtained after a shear treatment of about 1 MPa, one has an increase by about seven decades. Without any doubt this is a tremendous effect of shearing.

However, also another comparison must be made, namely with the total number of macromolecules per cubic meter. If as many as hundred macromolecules are assigned to every nucleus, one still finds that only one macromolecule is engaged in nuclei against a million of still unbound macromolecules. As all macromolecules, which are still unbound after the shear treatment, have undergone the same mechanical treatment, the conclusion must be that under usual conditions a successful association is an extremely rare event. This insight will be of importance for the interpretation of shear induced nucleation. In this respect, how-ever, there must be made a remark with respect to the temperature range, where the measurements of Fig. 1.1 have been made. In fact, a temperature range above 145 °C was chosen for the experiments in shear. The reason for this choice was that at lower temperatures, where nuclei come up also in the quiescent melt of it-PP, the initial conditions for shearing become quite uncertain. Also, with processing one always comes from higher temperatures. This means that the said temperature range above 145 °C must always be passed, leaving its marks also in the range of the low final temperatures.

1.2   Available Theories Describing the Crystallization Process

1.2.1   The Kolmogoroff-Avrami-Evans Theory

Kolmogoroff was the first to describe the consequences of nucleation and growth for the space covering by the formation of spherulites [18]. Later Avrami [19] and Evans [20] independently developed their theories, which meanwhile turned out to be essentially identical with Kolmogoroff's theory. Also Tobin [21] developed a theory, which he considered as an improvement of Avrami's theory. However, the results of this latter theory only deviate from those of the previous theories as a consequence of an unnecessary simplification introduced into the course of the calculation.

The goal of the present monograph is to deliver an introduction, which does not deter the reader by the use of advanced mathematics. Friends of such an approach are relegated to our review of 1997 [22] and to a later section of this book on confined samples. To start with, an equation is given, which relates the real space covering ξ_g to the virtual space covering ξ_g^*, which is obtained, if the impingement of the formed particles is disregarded. Subscript g stands for geometric. Both coverings are per unit volume. One has for the respective changes of these space coverings:

$$d\xi_g = \left(1 - \xi_g\right) d\xi_g^* \tag{1.1}$$

In fact, for the increase of the real space covering only the fraction $(1 - \xi_g)$ is available. This equation can easily be integrated. One obtains:

$$\xi_g = 1 - \exp(-\xi_g^*) \tag{1.2}$$

But this means that one has to calculate only the virtual space covering ξ_g^*. If spherulites are growing, one obviously has:

$$\xi_g^*(t) = \frac{4\pi}{3} \int\limits_{-\infty}^{t} du \, \alpha(u) \left[\int\limits_{u}^{t} dv \, G(v)\right]^3 \tag{1.3}$$

where α is the rate of formation of nuclei per unit volume and G is the growth speed of the spherulites. The expression between square brackets is the radius of a spherulite, which is born at time u, as observed at time t after unimpeded growth. If the expression of Eq. (1.3) is introduced into the exponent of Eq. (1.2), one obtains the famous equation, which has been given for the first time by Kolmogoroff. One has:

$$\xi_g(t) \;=\; 1 \;-\; \exp\left\{-\frac{4\pi}{3}\int_{-\infty}^{t} du\,\alpha(u)\left[\int_{u}^{t} dv\,G(v)\right]^{3}\right\}. \qquad (1.4)$$

The number density of nuclei at time t is then:

$$N_c(t) \;=\; \int_{-\infty}^{t} du\,\alpha(u)\left[1 - \xi_g(u)\right], \qquad (1.5)$$

where ξ_g is here the real space covering of Eq. (1.4). If there is a prescribed relation between temperature T and time t, namely $T(t)$, one can use the just given equations also outside the regime of isothermal processes. If the time dependence of α and G are only via the time dependent temperature, one has

$$\alpha(t) = \tilde{\alpha}(T(t)) \quad \text{and} \quad G(t) = \tilde{G}(T(t)). \qquad (1.6)$$

However, if the course of the (local) temperature is a consequence of heat transfer, there is no prescribed relation between temperature and time. As a consequence, the above equations are not directly useful under those conditions. It will be clear to the reader that the change of specific volume, which occurs with the crystallization, is disregarded in these equations. This fact is expressed by the choice of the name geometric space covering.

For isothermal crystallization only two cases are of importance, namely the case, where the nuclei are there from the beginning, and the case, where the nucleation rate is constant and finite. In the first case one has a fixed number density of athermal nuclei or of heterogeneous nuclei, the latter being caused by nucleation agents or impurities. In the second case the melt is clean, and sporadic nucleation of nuclei is occurring. The critical point in both cases is the choice of time zero. In fact, the melt, which is originally kept at a temperature well above the equilibrium melting point, must be quenched fast enough to the temperature, where the crystallization is intended to happen. The time interval needed for the quench must be very short compared with the time required for the crystallization process proper. Only under this condition time zero can readily be chosen within the time interval needed for the quench. In the case of sporadic nucleation the rate of formation of nuclei can be kept constant as long as the temperature is constant. In fact, this nucleation process occurs in the still uncovered volume of the sample.

For practical reasons it has been assumed [18] that originally there is a constant number density N_v of points, where nuclei can be born. If for these points an activation mechanism is assumed, one has for the number density $N(t)$ of already activated nuclei:

$$N(t) = N_V \left[1 - \exp\left(-\frac{t}{\tau}\right)\right]$$

where τ is an activation time. If this equation is differentiated, one obtains for the rate of formation of nuclei per unit volume:

$$\alpha(t) = \frac{N_V}{\tau} \exp\left(-\frac{t}{\tau}\right). \tag{1.7}$$

The two limiting cases, as mentioned above, are obtained for t much longer than τ and for t much shorter than τ. With $\alpha(t)$ according to Eq. (1.7) one can obtain single integrated versions of Eqs. (1.4) and (1.5), i.e.:

$$\xi_g(t) = 1 - \exp\left[-8\pi N_v\, G^3\, \tau^3\, e_4\left(-\frac{t}{\tau}\right)\right] \tag{1.8}$$

and

$$N(t) = N_v \int_0^{\frac{t}{\tau}} du\, \exp\left[-u - 8\pi N_v\, G^3\, \tau^3\, e_4(-u)\right], \tag{1.9}$$

where $e_4(z) = \sum_{i=4}^{\infty} \frac{z^i}{i!}$ is an exponential rest series.

For the important cases mentioned above, the direct integration of Eq. (1.4) is the easiest way. One obtains for $\alpha(t) = N_v\, \delta(t)$, where δ is the Dirac function concentrated on $t = 0$ ($\tau \to 0$):

$$\xi_g(t) = 1 - \exp\left[-\frac{4\pi}{3} N_v\, G^3\, t^3\right] \tag{1.10}$$

with $N = N_v$. The exponent of time is called the Avrami index, which is 3 for spherulitic growth, if the nuclei have been there from the beginning.

For $\alpha = \alpha_c = $ const. one finds:

$$\xi_g(t) = 1 - \exp\left[-\frac{\pi}{3} \alpha_c\, G^3\, t^4\right]. \tag{1.11}$$

In this case the Avrami index is equal to 4. The general form of these equations is:

$$\xi_g(t) = 1 - \exp(-k\, t^n),$$ (1.12)

with

$$k = \frac{4\pi}{3} N_V G^3 \quad ; \quad n = 3$$

and,

$$k = \frac{\pi}{3} \alpha_c G^3 \quad ; \quad n = 4.$$ (1.13)

As is well known, there are also two dimensional and one dimensional growth mechanisms. From a point of view of structure formation one has to consider only two cases: The growth from a thread-like precursor, which is created during a short period of heavy shearing or stretching and the growth from a flat surface, which is the wall of a container. If temperature is low enough, one has in the first case:

$$\xi_g(t) = 1 - \exp\left\{-2\pi L\left[\int_0^t dv\, G(v)\right]^2\right\} = \pi L\, G^2\, t^2$$ (1.14)

where L is the total length of thread-like nuclei per unit volume. This length can be estimated experimentally like the total length of wooden logs of thickness D and of 1 m length in a stack of 1 m^3. One has:

$$L = \frac{2}{\sqrt{3} \cdot D^2}.$$ (1.15)

For this purpose one has to estimate the average distance D between the threads from a microscopic picture. The reader may ask, why the growth mechanism of L is not introduced. The answer consists of two parts. First of all, the pertinent growth mechanism, which holds for the period of flow, has been a subject of discussion for many years. This subject can only be treated in Chap. 3, where the influence of flow is examined. Second, it has not yet been possible to determine the growth speed G of cylindrites. For the melt, which is quieted down after the period of flow, one can assume that this growth speed is equal to the relatively slow growth speed characteristic for spherulites in a quiescent melt. Equation (1.14) is valuable, because many useful experiments have been carried out so far, obeying the conditions, under which this equation can be used. In many industrial processes short term flow is applied (see injection molding). Our experiences have shown that the typical structures grow quite slowly only after the cessation of the flow, even if they differ from those, which grow in permanently quiescent melts.

The mentioned one-dimensional growth occurs with transcrystallization [23]. It also influences the result of differential scanning calorimetry (DSC) at temperatures not too far from the equilibrium melting point, where the number of nuclei in the melt is extremely low and, as a consequence, nucleation by the wall dominates. One has:

$$\xi_g(t) = A \int_0^t dv\, G(v) = A\,G\,t, \tag{1.16}$$

where A is the surface area in front of the unit volume. One can clearly see from Eqs. (1.14) and (1.16) that the Avrami index is 2 in the two dimensional case and 1 in the one dimensional case. The equations, which contain an integral over the time, are also valid for a growth speed depending uniquely on the temperature. The equations containing G as such, hold only for the isothermal cases. For these latter cases the so-called Avrami plot has been invented. It reads:

$$\log(-\ln(1 - \xi_g)) = n\,\log(t) + \log(k). \tag{1.17}$$

This equation gives a linear plot with respect to $\log(t)$ with n as the slope. This plot became very popular in the past. However, the points of this plot have very different weights depending on their location on the graph. This fact makes the position of a straight line, as drawn through the experimental points, very unsecured. The position of this line is extremely sensitive to errors in small values of ξ_g, for which no secure determination is thinkable, and for the moment of zero time. In fact, the value of ξ_g depends on the specific volume, which deviates only little from that of a melt still containing no crystalline domains. Light scattering increases in the beginning only with the sixth power of time. Small values of latent heat cannot easily be detected. And also $\log(t)$ is uncertain. What is the moment, when a melt, which must be cooled down from a temperature well above the equilibrium melting point, reaches the intended crystallization time? For all these doubts the reader is referred to a critical paper by Eder [24]. So, it does not surprise, if all kinds of Avrami indices, which do not agree with reasonable expectations, have been reported. And also, if DSC is used, the trans-crystallization effect can reduce the effective Avrami index, which otherwise will, for instance, be equal to 3.

An ingenious idea for coping with the influence of temperature, if the course of this temperature cannot be prescribed, has been published by Nakamura, Watanabe, Katayama and Amano almost thirty years ago [25]. For the purpose these authors introduced a dimensionless time replacing the quotient t/τ in Eqs. (1.6) and (1.7). The latter equation reads now:

$$\alpha(t) = \frac{N_v}{\tau(t)} \exp\left[-\int_0^t \frac{du}{\tau(u)}\right]. \tag{1.18}$$

The extra condition, which is introduced here, is the so-called isokinetic assumption, which means that the growth speed and the activation frequency of primary nuclei $1/\tau(t)$ have the same temperature dependence. Obviously, for this assumption the activation process of nuclei, as has been sketched by Eq. (1.6), is of importance. It means that one has:

$$G(t)\,\tau(t) = \text{const.} = \lambda. \tag{1.19}$$

In this equation λ is a characteristic length, which cannot have a completely arbitrary value. In fact, the progress in growth is restricted by the time, which elapses up to the birth of the next nucleus. If the product $G^3\tau^3$ in Eqs. (1.8) and (1.9) is simply replaced by λ^3, one obtains the corresponding equations for the case of isokinetics. In these equations the ratio t/τ must be replaced by

$$w(t) = \int_0^t \frac{du}{\tau(u)}. \tag{1.20}$$

The lower bound zero of the integral means that time zero is at the moment, when during cooling the sample passes the equilibrium melting point. In fact, τ can be assumed to be infinite above this temperature. One has:

$$\xi_g(t) = 1 - \exp\left[-\,8\pi\,N_v\,\lambda^3\,e_4\left(-w(t)\right)\right] \tag{1.21}$$

and

$$N_c(t) = N_v \int_0^{w(t)} dw\,\exp\left[-w - 8\pi\,N_v\,\lambda^3\,e_4(-w)\right]. \tag{1.22}$$

In these equations the thermal history of the cooling process, which normally varies in the sample from place to place, is represented by the value of $w(t)$. In particular, this function shows up as the upper bound of the integral of Eq. (1.22).

A criterion for the quality of the description is the number density of spherulites at infinite time, as this density is characteristic for the obtained structure of the sample. Since also $w(t)$ goes to infinity with t, one has:

$$N_c(\infty) = N_v \int_0^{\infty} dw\,\exp\left[-w - 8\,\pi\,N_v\,\lambda^3\,e_4(-w)\right]. \tag{1.23}$$

Unfortunately it must be said that, obviously, this integral is a constant independent of the thermal history. But this means that the isokinetic approach is unable to predict a real structure. The structure, as obtained according to this approach, is predetermined by the chosen values of N_v and λ [22]. The general experience that the number density of spherulites increases strongly with the cooling speed, is not reflected by Eq. (1.23). As a consequence, also local variations of the structure, as found in every sample, cannot be described. It can also happen that $w(t)$ does not go to infinity because of the fact that the glass transition temperature is passed before crystallization is complete. In this case only a number density of spherulites lower than the one predicted by Eq. (1.23), can be expected according to this approach.

It also remains questionable, whether the evolved heat of crystallization is correctly described, if these equations are applied. In fact, this heat of crystallization interferes, dependent on the location in the sample, with the solution of the heat transfer problem. Because of the fact that, apparently, the isokinetic approach is not realistic, an almost thirty years' history of success, which has been booked for the equations of Nakamura et al., must be ended. Meanwhile, there is also direct evidence for the inadequacy of the isokinetic approach. In fact, if one looks at the experimental results shown in Sect. 1.1, one notices that the number density of nuclei increases continuously with decreasing temperature, whereas the growth speed goes through a maximum halfway down to the glass transition temperature. As will be shown later in this monograph, these experiences hold for a great number of polymers.

The conclusion must be that for a correct treatment of non-isothermal processes a more powerful method must be found. Such a method will be described in the next section.

1.2.2   The Rate Equations of Schneider, Köppl and Berger

A combination of Kolmogoroff's integral, as given in Eq. (1.3), with the equation of heat conduction, which is a differential equation, does not seem an easy task. As a consequence, Schneider et al. [26] decided to transform this integral into a more suitable configuration. For the purpose they differentiated this integral several times with respect to time. It appeared that with every step a useful auxiliary function was obtained. In this way a system of differential equations was created. The authors called these differential equations the rate equations. If the right side of Eq. (1.3) is denoted as φ_0, this system of equations reads in a slightly modified version:

$$\frac{d\varphi_i(t)}{dt} = \frac{1}{G(t)}\,\varphi_{i-1}(t).$$

(1.24)

for $i = 1, 2, 3$ and $\varphi_i(-\infty) = 0$. The latter condition just means that the history of the crystallization process is considered from a moment long before crystallization started. In Eq. (1.20) the begin was set at time zero for the same purpose. To start with one has:

$$\varphi_0(t) = \frac{4\pi}{3} \int\limits_{-\infty}^{t} du\, \alpha(u) \left[\int\limits_{u}^{t} dv\, G(v) \right]^3, \tag{1.25}$$

where φ_0 is identical with ξ_g^* of Eq. (1.3). It is the total volume of spherulites per unit volume of the sample, if the impingement (space filling effect) is disregarded. In fact, this so-called undisturbed volume can formally become larger than the unit of volume of the sample. But this does not matter, as the necessary correction, which is given by Eq. (1.2), can always be carried out. The next auxiliary function is:

$$\varphi_1(t) = 4\pi \int\limits_{-\infty}^{t} du\, \alpha(u) \left[\int\limits_{u}^{t} dv\, G(v) \right]^2 \tag{1.26}$$

This φ_1 is the total surface area of the undisturbed spherulites per unit volume of the sample. This function is of particular interest because the growth occurs on this surface. In continuation one has:

$$\varphi_2(t) = 8\pi \int\limits_{-\infty}^{t} du\, \alpha(u) \left[\int\limits_{u}^{t} dv\, G(v) \right], \tag{1.27}$$

which is 8π times the sum of the radii of the undisturbed spherical particles per unit volume, and

$$\varphi_3(t) = 8\pi \int\limits_{-\infty}^{t} du\, \alpha(u), \tag{1.28}$$

which is 8π times the number of these particles per unit volume. If one is interested in the rate $\alpha(t)$ of nucleation, one also can differentiate Eq. (1.28). One obtains:

$$\frac{d\varphi_3(t)}{dt} = 8\pi\, \alpha(t). \tag{1.29}$$

For reasons already mentioned, the first of the Eqs. (1.24) is of particular interest. After a rearrangement it reads:

$$\frac{d\varphi_0(t)}{dt} = G(t)\, \varphi_1(t). \tag{1.30}$$

This equation expresses the fact that growth occurs at the surface of the particles. If impingement is taken into account, one has:

$$\frac{d\xi_g(t)}{dt} = G(t)\,\varphi_1(t)\exp(-\varphi_0(t)) \tag{1.31}$$

This equation is directly obtained by a differentiation of Eq. (1.4). If secondary crystallization is disregarded in a fast cooling process, the real degree of crystallinity ξ becomes:

$$\xi = p\xi_g, \tag{1.32}$$

where p (<1) is the fraction of crystallinity shortly after the manufacture.

For the moment only the one dimensional case of heat transfer will be mentioned. For this case the equation of heat conduction can be written as follows:

$$\frac{\partial T}{\partial t} = a\frac{\partial^2 T}{\partial x^2} + \frac{h}{c}\frac{\partial \xi}{\partial t}. \tag{1.33}$$

In this equation a is the heat diffusivity of the melt, h is the specific heat of crystallization and c the specific heat capacity per unit mass. It is now evident, how the equations are coupled by the evolved heat of crystallization.

The rate equations form the basis for the pertinent calculations. However, without knowledge of the time dependence of the kinetic parameters G and α these equations cannot become operative. As already mentioned in Sect. 1.1, G can be considered as not directly dependent on time. The situation with α is, in principle, a little more complicated. However, as the experimental results of Fig. 1.1 suggest, a unique function of temperature can be assumed for the number density N of the nuclei after isothermal crystallization. But this means that α can be expressed as:

$$\alpha(t) = \frac{dN(T(t))}{dt} = \frac{dN(T(t))}{dT}\frac{dT(t)}{dt} \tag{1.34}$$

It should be observed that both differential quotients on the right side of this equation carry a negative sign for α being positive with cooling.

The number density of nuclei as a unique function of temperature has previously been proposed by Van Krevelen [27] as the most practical approximation. However, it will be shown in Chap. 2 of this monograph that the overwhelming majority of nuclei in polymer melts is actually of the athermal type. For those nuclei the unique function of temperature is realistic.

It turns also out that Eq. (1.34) must be modified for the case that, as a consequence of the evolution of latent heat, local re-heating occurs in the sample. But it is evident that nuclei, which come up during cooling, immediately experience some stabilization, so that they survive a modest re-heating. In taking account of this effect, the course of the real local temperatures T must be modified for the time span, where the local reheating occurs. During this time span the real temperature must be replaced by the lowest temperature $T_{min}(x,t)$. This temperature is the temperature of the minimum, where the upturn sets in. Until with further outside cooling the real temperature returns to the value, which it had at the upturn, this T_{min} will be kept constant. During this time span no new nuclei should come up.

For the development of the morphology Eq. (1.34) can be used, if for the term between brackets $\exp(-\varphi_0)$ is inserted (cf. Eqs. (1.4), (1.5) and (1.25)):

$$\frac{dN_c}{dt}(t) = \frac{dN}{dT}(T(t)) \frac{dT}{dt}(t) \exp(-\varphi_0(t)). \tag{1.35}$$

This equation can be integrated simultaneously with the Eqs. (1.24), (1.28) and (1.33). This integration starts, before the temperature at the surface of the sample passes the equilibrium melting point during the process of cooling.

However, in principle there is still a problem of a different nature. A straight forward use of Eq. (1.33) is only possible, if the temperature gradients in the sample remain small enough for the assumption that every newly formed spherulite can simply grow at the calculated temperature of its nucleus. In fact, if the temperature shows a noticeable gradient already within the domain of the growing spherulite, this spherulite will show a faster growth on the side of the lower temperature. A description of this more complicated situation must be postponed to a later section. For the fine grained structures, which are desirable from the point of view of technology, such an improvement will probably not be necessary.

Another point is, that in practical processing one can certainly not disregard the influence of flow. But this means that in first instance one has to add a convection term to Eq. (1.33). In this way Eq. (1.33) is transformed into the so-called energy equation. Fortunately, one does not need the general case of this equation. In all relevant cases the fluid flows with high speed along a sufficiently flat wall of lower temperature. For those cases one has:

$$\frac{\partial T}{\partial t} = a \frac{\partial^2 T}{\partial x^2} + \frac{h}{c} \frac{\partial \xi}{\partial t} + v_y(x) \frac{\partial T}{\partial y}, \tag{1.36}$$

where v_y is the fluid velocity in the y-direction as a function of distance x from the wall. For this equation it is assumed that heat conduction can be neglected with respect to heat convection in the y-direction. Only in the x-direction, which is perpendicular to the wall, conduction is of great importance. Nevertheless, a direct use of this equation cannot achieve its object. In this respect it is not the problem of an accurate calculation of the convection term, even if the rheology of the fluid is modified by incipient crystallization. The difficulty lies mainly in the second term on the right side. In fact, proper experiments have shown that the kinetics of crystallization are extremely sensitive to velocity gradients. And for continued flow this problem has not yet been settled. This fact will be discussed in Chap. 3 of this monograph.

Interestingly, for a number of practical situations there is an escape from this dilemma. In fact, it turned out that in short term experiments, when flow was applied only for rather short periods, number and character of the nuclei changed tremendously. But growth on these nuclei seemed to be delayed. In fact, after cessation of flow it took considerable time spans, before growth became noticeable. Apparently, the surface area of the newly formed nuclei, in particular of those, which are very thin threads, is extremely small in the beginning. In the quieted down melts it took waiting times, which were a hundred or even a thousand fold longer than the flow times. Only after these waiting times crystallization became noticeable by light scattering or depolarization of polarized light. The increased number densities of nuclei, as given on the right side of Fig. 1.1, were obtained from the number densities of particles finally found. However, quite recently also long term shearing has been introduced. The consequences of this treatment, however, will be discussed, when the results of this treatment will be presented.

For the preliminary use of Eq. (1.36), however, the convection term remains unchanged with respect to a melt, which did not undergo a crystallization process. And also, no considerable amount of latent heat h is evolved during the period of flow. In particular, the interpretation of structures in injection molded parts can be facilitated by these facts.

1.3   Examples for Special Cases

1.3.1   Minimum Cooling Speed for by-Passing Crystallization, Eder [28]

If the cooling speed is prescribed, one has a case, where Kolmogoroff's Eq. (1.4) can directly be applied without a recurrence to the troubles caused by the solution of a heat transfer problem. In fact, the heat of fusion h does not play a role, if the cooling speed is so fast that crystallization does not get a chance (see the title of this section). However, a constant cooling speed can be achieved for the interior of a still amorphous sample, if the coolant has a temperature, which decreases linearly with time and if the sample itself is sufficiently thin. A rule of thumb says that the Fourier number must be smaller than 0.5 for the achievement of steady cooling. For a cylindrical sample this means:

$$Fo = \frac{a\,t}{D^2} < 0.5, \qquad (1.37)$$

where D is the sample diameter and t is the running time, which is required for an achievement of steady cooling. This equation shows that the running time t decreases with the square of the sample diameter. The fact that a finite running time is unavoidable means that the temperature, at which the sample is exposed for the first to the coolant, must be higher than the temperature, where crystallization can be expected to start with slower cooling. One has for the pertinent temperature difference:

$$\Delta T \approx q\, t, \tag{1.38}$$

where q is the constant cooling speed of the coolant, as governed by the following equation:

$$T(t) = T_{\max} - q\, t \tag{1.39}$$

For convenience the reference temperature T_{max} is chosen as the temperature, where the growth speed $G(T)$ of the spherulites shows its maximum (cf. Fig. 1.2).

For the following calculations quite realistic approximate equations are chosen for $G(T)$ and $N(T)$. These equations read:

$$G(T) = G_{\max} \exp\left(-\left(k_G(T - T_{\max})\right)^2\right) \tag{1.40}$$

and

$$N(T) = N_{\max} \exp\left(-k_N(T - T_{\max})\right) \tag{1.41}$$

In fact, the temperature dependencies of these parameters must be known, if their influences are to be minimized. One can observe that the equation for the number density does not show a maximum. The symbol N_{max} stands for the number density at the temperature T_{max}, where the growth speed has its maximum. For the calculations one needs the rate of nucleation. For N as a unique function of temperature one obtains this rate from Eq. (1.34), where the term between brackets is omitted for the present purpose. In the present case the rate of nucleation, which is undisturbed by the impingement effect, is needed, as the degree of crystallinity remains negligible. The function φ_0, as occurring in the exponent of Kolmogoroff's equation, is required. If Eq. (1.41) is differentiated with respect to T, and since dT/dt is equal to $-q$, one has:

$$\alpha(t) = N_{\max} k_N\, q\, \exp(k_N q\, t) \tag{1.42}$$

Equation (1.25) can now be integrated, if in Eq. (1.40) $k_G(T - T_{max})$ is replaced by $-k_G\, q\, t$ (according to Eqs. (1.39)) and (1.42) is used as it stands. One obtains in this way:

$$\varphi_0(t) = K f(k_G q t\,;\, k_N/k_G). \tag{1.43}$$

Explicitly function f reads:

$$f(v\,;\, \kappa) = \frac{\kappa}{8} \int_{-\infty}^{v} ds\, \left[\operatorname{erf}(v) - \operatorname{erf}(s)\right]^3 \exp(\kappa s) \tag{1.43a}$$

with $v = k_G\, q\, t$ and $\kappa = k_N/k_G$. The constant K reads:

$$K = \frac{4\,\pi^{5/2}}{3}\,\frac{N_{\max}\,G_{\max}^3}{k_G^3}\,\frac{1}{q^3} \tag{1.43b}$$

The critical cooling speed for by-passing crystallization is defined here as that cooling rate q_{crit}, giving at the end of the cooling ($t = \infty$) a degree of space covering of ξ_g of 0.01, as a space covering of zero would require an infinitely fast cooling speed. At this low space covering ξ_g and φ_0 are identical. So one has:

$$q_{crit} = \sqrt[3]{\frac{400}{3}}\,\pi^{5/6}\,\frac{G_{\max}\,\sqrt[3]{N_{\max}}}{k_G}\,\sqrt[3]{f(\infty\,;\,k_N/k_G)} \tag{1.44}$$

It turns out that the numerical value of $f(\infty;\kappa)$ varies between 1 and 0.6 in the relevant range of κ between zero and five with its minimum of ≈ 0.6 at $\kappa \approx 3$. As the third root of f occurs in Eq. (1.44), the last term in Eq. (1.44) deviates at most by 0.15 from 1. Ignoring this maximum deviation one obtains:

$$q_{crit} = 13.24\,\frac{G_{\max}\,\sqrt[3]{N_{\max}}}{k_G} \tag{1.45}$$

At this point one would like to find a table, which contains the critical cooling speeds q_{crit} for a series of industrially important polymers, as an illustration of the importance of Eq. (1.45). Notwithstanding the fact that the required kinetic data G_{max}, N_{max} and k_G can only be scrutinized later in the context of their determination (see Chap. 2 of this monograph), such a table will be given here as Table 1.1.

From this table one can learn that critical cooling speeds vary by six decades from HDPE to i-PS. It will be seen later that the behavior of crystallizing polymers fills the gap between the behavior of metals and of glass forming minerals.

Probably, a few words must be said about the character of the approximate Eq. (1.40). In principle, the growth speeds go to zero at the melting point T_m and at

Table 1.1 Minimum cooling speeds for bypassing crystallization in a series of quiescent industrial polymers, with required kinetic data

Polymer	G_{max} [ms^{-1}]	k_G [K^{-1}]	N_{max} [m^{-3}]	q_{crit} [Ks^{-1}]	Grade
HDPE	$>8 \times 10^{-4}$	~ 0.05	$>10^{17}$	**>10000**	Borealis
PK	9.7×10^{-6}	0.036	3×10^{14}	**620**	RDP-211 Shell
	5.4×10^{-6}	0.045	6×10^{13}	**240**	Carillon Shell
iPP	5.0×10^{-6}	0.051	10^{14}	**60**	KS10 Borealis
PB	1.6×10^{-6}	0.072	6×10^{14}	**25**	0110 Shell
PET	2.5×10^{-7}	0.033	1.5×10^{15}	**11.5**	MPET Sinco
	4.0×10^{-7}	0.034	1.5×10^{14}	**8.3**	DMT Sinco
	2.0×10^{-8}	0.030	10^{13}	**~0.9**	v. Antwerpen
iPS	2.5×10^{-9}	0.035		**~0.02**	v. Krevelen

the glass transition temperature T_g. In fact, at temperatures above the equilibrium melting point there are no (secondary) nuclei promoting growth and below the glass transition temperature the extremely high viscosity is prohibitive of any transport mechanism. However, these facts are not reflected by Eq. (1.40). This equation does not contain the parameters T_m and T_g. However, in this equation the growth speeds seem to become low enough on both sides of T_{max} for being of no considerable influence on the integration. Also the fact that with polymers of decreasing crystallization speed the distance between the melting point and the glass transition temperature decreases continuously, is practically taken into account by this equation. In fact, if G_{max} is low, one quickly arrives on both sides of T_{max} at sufficiently low values of the growth speed.

1.3.2 A Dimensionless Process Classification Number

The idea for the definition of such a number goes back to 1984 [29]. This number can be defined as:

$$Jk = \frac{\tau_{th}}{\tau_{cr}}, \tag{1.46}$$

where τ_{th} is the time needed for thermal equilibration (termination of the cooling process) and τ_{cr} is the time needed by the crystallization process. The symbol Jk has its origin in a paper by Astarita and Kenny [30]. These authors became fond of the idea and called the number the Janeschitz-Kriegl number. It turns out that it is not difficult to find a classical expression for τ_{th}. It reads:

$$\tau_{th} = \frac{(1 + Ste)\, d^2}{a} \quad \text{with} \quad Ste = \frac{\Delta h}{c_p\left(T_f - T_s\right)} \tag{1.47}$$

In this equation d is the thickness of the sample (a slab), a is the heat diffusivity of the sample, as averaged over the states of aggregation and Ste is the Stefan number with Δh being the specific latent heat corresponding with the final degree of crystallinity, c_p the averaged specific heat, T_f the starting temperature of the fluid and T_s the final temperature of the solid.

Unfortunately, we were not so successful in the formulation of the time τ_{cr}, which is required for the crystallization process. As a consequence it took quite a time, until Jk could be made operational. In fact, in the beginning we did not have the required kinetic data at our disposal. Nevertheless we were able to show that the metals were on one side of the spectrum and the glass forming minerals on the other side of the spectrum with the polymers in between. This comprehension was not so difficult to achieve, because of the enormous differences in the crystallization speeds.

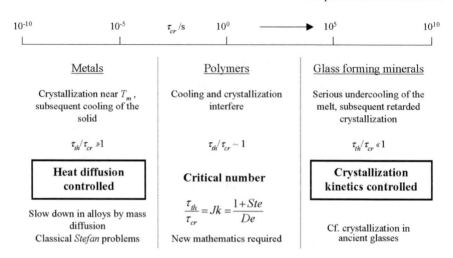

Fig. 1.4 A viable classification of materials on the basis of their processing conditions

These speeds cover about 16 decades, whereas the heat diffusivities of the diverse materials cover only about 4 decades. As a consequence, manufacturing processes of metals with their enormous heat conductivities nevertheless are of the heat diffusion type. In fact, it is always the slower partial process, which determines the character of the whole process. But crystallization is so fast with metals that it always wins against cooling. Only with extremely thin samples one can find solidification processes, which are crystallization kinetics controlled, where crystallization becomes the slower process. The situation is completely different with glass forming minerals. With these minerals crystallization is extremely sluggish. As a consequence, even very thick samples still show crystallization kinetics controlled crystallization. The cooling speed is always too large for a heat diffusion controlled process. The outcome of this consideration is depicted in Fig. 1.4. It gives a viable classification of materials [31].

On top of this compilation one finds the scale of the crystallization times τ_{cr}. A time scale for the heat diffusivity is not shown. Its influence is secondary. In the middle one finds the Jk number. In this equation one finds the Deborah number $De = a\tau_{cr}/d^2$. But for polymers this presentation is too coarse. For these materials one needs an explicit expression for the time τ_{cr}.

From the consideration in the previous section we take over the suggestion that the product of $G_{max}N_{max}^{1/3}$ will be representative. If a sample is quenched fast enough to-say-T_{max}, all spherulites start growing simultaneously on the number of nuclei representative for this temperature.

In this case one obtains the following form of Kolmogoroff's equation (see Eq. (1.10)):

$$\xi_g(t) = 1 - \exp\left(-\frac{4\pi}{3} N_{max} G_{max}^3 t^3\right) \tag{1.48}$$

According to this equation space covering is sufficient, if the argument of the exponential is -1. In this way the crystallization time can be calculated for this situation. One has:

$$\tau_{cr} = \frac{1}{1.63\ G_{max} N_{max}^{1/3}} \tag{1.49}$$

If this expression is used together with Eq. (1.47) in Eq. (1.46), one has:

$$Jk = \frac{(1 + Ste)\,d^2}{a}\ 1.63\ G_{max} N_{max}^{1/3}. \tag{1.50}$$

This equation seems very acceptable. Nevertheless, some discussion seems necessary.

For the purpose the following train of thoughts is proposed: As a first step the fluid sample is cooled to T_{max} with the minimum speed, which prevents crystallization. In a second step the temperature is kept at T_{max}, until the sample has completely crystallized according to the criterion of Eq. (1.49).

With the said cooling step only about half the distance between T_m and T_g is covered. This means that one needs only half the time, which is calculated from Eq. (1.45). In this way one has $t_{cool} = (T_m - T_g)/(2q_{crit})$.

According to the Gaussian curve for $G(T)$ one has $k_g = 2/(T_m - T_g)$. A combination of these expressions with Eq. (1.45) yields:

$$t_{cool} = \frac{1}{13.24\ G_{max} N_{max}^{1/3}}. \tag{1.51}$$

As this time is only about one eighth of the time needed for the crystallization according to Eq. (1.49), this time t_{cool} can be disregarded in the rough estimate. But this means that for the polymers of medium speed of crystallization Eq. (1.50) seems quite useful. As a consequence, this equation is used for the calculation of the Jk numbers of the polymers, which were quoted already in Table 1.1. In this way Table 1.2 is prepared.

Table 1.2 also gives the required thermodynamic and statistical parameters. A sample thickness of 1 mm is assumed. With PK (polycarbonate) two copolymers were investigated. As no physical data were available, the value of τ_{th} was just estimated. There were three samples of PET of different molar masses. For this polymer the growth speeds depend on the molar mass, as the chain lengths are rather short in general. This fact has been documented by Van Antwerpen and Van Krevelen [32].

Table 1.2 Thermodynamic and kinetic data leading to a calculation of the Jk number. Courtesy of Hanser Verlag [31]

	ΔH	T_m	Ste	C_p	a	τ_{th}	$1/\tau_{cr}$	Jk
	[kJ/ mol]	[K]	[l]	[J/ molK]	$[10^{-7}$ ms$^{-1}]$	[s] (d = 1 mm)	[s^{-1}]	[l] (d = 1 mm)
HDPE	8.22	415	1.12	63	1.3	16.3	561	9120
PK						≈20	1.04 0.34	20.8 6.50
iPP	8.70	483 (α–mod)	0.39	93	0.95	14.6	0.374	5.45
PB-1	7.00	411 (form I)	0.52	120	0.90	16.9	0.218	3.69
PET	26.9	550	3.58	298	1.43	32.0	4.76×10^{-2} 3.51×10^{-2} 0.32×10^{-2}	1.52 1.12 0.103
i-PS	10.0	516	2.58	178	1.0	35.8	8.71×10^{-5}	0.0031

One notices that for the investigated polymers the values of *Jk* spread over many decades like the values of the critical cooling speeds (see Table 1.1). The difference is only that in Table 1.2 also the sample thickness plays a role. If instead of 1 mm a thickness of 0.33 mm is introduced, all values of *Jk* are reduced by a factor ten.

If one now looks at the edges of the spectrum, one can gain further insights. With fast crystallizing polymers like HDPE cooling of the samples cannot be fast enough for avoiding premature crystallization. The sample starts to crystallize much too early. The evolved latent heat causes a re-heating, so that the samples crystallize close to the melting point. The time of crystallization is increased in this way with respect to the value calculated with Eq. (1.49). But this means that the *Jk* number is reduced. If a thinner sample is used, this effect becomes less pronounced.

If one now looks on the other edge of the spectrum, one can give the following analysis: With i-PS the minimum cooling speed for preventing crystallization is very low. So it takes hours, before T_{max} is reached, if the said concept is respected. Everyone tends to use too high a cooling speed. But in such a case, the glass transition temperature is reached before the onset of a one percent crystallization, as is required by the concept of Eder [28]. One obtains a sample with a much lower degree of crystallinity as an end product.

The enormous differences in crystallization speeds, as documented in the Tables 1.1 and 1.2 necessarily provoke a serious discussion. For the purpose three polymers with a C-C-back-bone are chosen, namely linear high density polyethylene, isotactic polypropylene and isotactic polystyrene. The crystallization speeds of HDPE and it-PS differ by a factor of one million. The speed of it-PP lies in between. Without any doubt these differences have their origin in the kinetics. In fact, the melting point of it-PS is higher than that of it-PP, which on its part is higher than that of PE. This fact means that the most stable crystal is most seriously hampered in its formation. Apparently the probability that the conformations fit at the spots of contact, is a question of the back-bone structure. Nevertheless, the enormous differences remain puzzling. It seems that the only explanation lies in the fact that one has to do with many particles problems. In fact, the contact between two molecules is insufficient for the formation of a nucleus. So, one has to do with the multiplication of a number of pairing probabilities, each of them being smaller than one. In fact, parts of several molecules must fit simultaneously. But by this multiplication the probability for the formation of a nucleus must become very low and the differences of crystallization speeds are augmented. A mitigation of this harsh condition can only be achieved by a preceding arrangement of local alignments. This problem will keep us busy also with shear induced nucleation.

1.3.3 Scanning Calorimetry

Differential scanning calorimetry is a well-known technique for the evaluation of thermal properties of materials. As a consequence it is not surprising that this technique has also been applied to polymers in order to get an impression of their crystallization kinetics [33]. However, polymers are materials with very low heat conductivities, a fact which has not been taken to heart sufficiently. As a consequence, machines were usually calibrated only with respect to temperature with a number of suitable metals. A calibration with respect to heat transfer had not been carried out up to the moment when we started our critical investigations [34, 35]. In fact, metals have heat conductivities, which are factors of more than hundred larger than those of polymers. This means that the speed of cooling must have a tremendous effect with polymers, when compared with metals.

1.3.3.1 Unabridged Simulation

C.H. Wu successfully calculated the behavior of an industrial PP, of which except for the thermal data (heat conductivity, heat of fusion, final degree of crystallinity, specific heat and density) also crystallization kinetics data (number densities and growth speeds of spherulites as functions of temperature) were available to a sufficient extent [35]. Admittedly, at that time reliable growth speeds for the relevant α-modification of this polymer were known only for temperatures higher than 110 °C. An extrapolation to lower temperatures was carried out on the logarithmic scale of G with the aid of a tangent drawn to the part of Fig. 1.2, which existed already at that time (see Ref. [36], Fig. 1.7, open symbols, closed symbols holding for the β-modification). A similar, less critical extrapolation was carried out also for the number density of nuclei. In fact, this number density shows up as a first power in the equations, in contrast to the third power of the growth speed. Also a heat transfer coefficient, as determined for the transfer between pan and furnace with the aid of metallic samples [34] was at our disposition. As we are convinced that the general character of the results is not seriously influenced by the provisional kinetic low temperature data, characteristic results of these calculations are reproduced here. In these calculations the rate equations by Schneider et al. [26] were successfully used for the first time. Mrs. Wu, as a post doctorate fellow in mathematics, was just the right person for these involved calculations.

Reality lies between two limiting cases: The sample as a tablet is sandwiched between two metal walls (symmetric case) of equal temperature or the sample is in contact only with one metal surface. Its other surface is a free adiabatic surface (asymmetric case). In the DSC-machine a situation prevails, which just lies between the said limiting cases. The sample is in contact with the bottom of the pan which, on its part, is in touch with the furnace. On the upper side heat transfer is not directly zero but much less than at the bottom. The said limiting cases have the advantage that they can be treated together. In fact, at the mid-plane of the symmetric case the temperature gradient is zero. But the same holds for the adiabatic free surface of the asymmetric case. But this means that the solution for the asymmetric case is the same as for the symmetric case with doubled sample thickness. If in the latter case the sample formally has twice the thickness, one also must use twice the heat transfer coefficient on the contact surface.

The boundary value problem for the symmetric case can now be formulated. For the sample one has:

$$
\begin{aligned}
\frac{\partial T_s}{\partial t} &= a_s \frac{\partial^2 T_s}{\partial x^2} + \frac{h_s}{c_s} \frac{\partial \xi}{\partial t} &\quad \text{for all } (x,t) \in (D_p, D_p + D_s/2) \times (0, \infty) \\
T_s|_{t=0} &= T_i(x) = T_{rf} &\quad \text{for all } x \in \left[D_p, D_p + D_s/2\right] \\
\lambda_s \frac{\partial T_s}{\partial x}\Big|_{x=D_p} &= \lambda_p \frac{\partial T_p}{\partial x}\Big|_{x=D_p} &\quad \text{for all } t > 0 \\
\frac{\partial T_s}{\partial x}\Big|_{x=D_p + \frac{D_s}{2}} &= 0 &\quad \text{for all } t > 0
\end{aligned}
\tag{1.52}
$$

The symbols are T for temperature, D for thickness, a for heat diffusivity, λ for heat conductivity, h for the latent heat (corresponding to the attainable degree of crystallinity), c for specific heat, ξ for fraction of crystallized material, t for time and x for the distance from the surface of the furnace. Subscripts are p for pan, s for sample, i for initial and rf for reference. One finds the symbol T_{rf} in the equation describing the course of the furnace temperature T_f with time:

$$
T_f(t) = T_{rf} - qt \quad \text{with} \quad q = -\frac{dT_f}{dt} = \text{const.,}
\tag{1.53}
$$

which is the nominal constant rate of temperature change, as programmed for the furnace. The furnace is assumed to follow this temperature course exactly. For the interior of the wall of the pan one has:

$$\frac{\partial T_p}{\partial t} = a_p \frac{\partial^2 T_p}{\partial x^2} \qquad \text{for all } (x,t) \in (0, D_p) \times (0, \infty)$$

$$T_p\big|_{t=0} = T_i(x) = T_{rf} \qquad \text{for all } x \in [0, D_p] \qquad (1.54)$$

$$\lambda_p S_1 \frac{\partial T_p}{\partial x}\Big|_{x=0} = \frac{\gamma}{2}\left(T_p\big|_{x=0} - T_f(t)\right) \quad \text{for all } t > 0$$

A few additional symbols are S_1 for one of the contact surfaces on one side of the sandwich and γ for the heat transfer coefficient between pan and furnace. For the symmetric case the total contact surface is $S = 2 S_1$. The heat transfer coefficient is subdivided into two equal parts $\gamma/2$ for each contact surface.

As one can see on Eqs. 1.52, 1.53 and 1.54, the values of an enormous number of parameters are required. Most of them, in particular the thermal data, can be found in the usual tables for polymers. In the original paper [35] there is an extended table of all necessary data. Here we content ourselves with referring to Figs. 1.1 and 1.2 of this report, from where the kinetic data can be extracted in principle. In fact, these data were needed in the rate equations [26] of Sect. 1.2.2, which were integrated by Mrs. Wu together with Eqs. 1.52 and 1.54. In addition a value for the heat transfer coefficient between furnace and pan was needed. This coefficient had been determined in Ref. [34] to $\gamma = 0.016$ W/K. The way, in which this coefficient has been obtained, will be described below in connection with the difference, which is made between metallic and polymer samples.

The results of the said calculations are given in Figs. 1.5 and 1.6. In these figures the apparent latent heat $l(T_f)$ is plotted for two samples of differing thickness (0.3 and 0.9 mm) against the furnace temperature for eight nominal cooling rates of 50, 40, 30, 20, 15, 10, 5 and 1 K/min. The apparent latent heat $l(t)$, which is considered as the DSC-signal, is defined as:

$$l(t) = \frac{\gamma\left(T_p\big|_{x=0} - T_b(t)\right)}{m_s q}, \qquad (1.55)$$

where T_b is the so-called "base line" (see later). Time t and furnace temperature T_f are related by Eq. (1.53).

Figures 1.5 and 1.6 show a similar tendency: With increasing cooling rate the peaks shift to lower temperatures and become lower and broader. For the thicker sample (Fig. 1.6) this effect is more pronounced.

The results of these calculations are compelling. In fact, if the heat transfer problem is ignored, the peaks have all the same height and shape. They are only shifted to lower temperatures with increasing cooling speeds [35, 36]. There is also another important result, which cannot be seen on these figures: With increasing cooling speed the temperature difference between the mid-plane of the sample and the furnace increases considerably, up to more than 12° centigrade even with the thinner sample. By the way, with the more realistic course of G as a function of temperature, as shown in Fig. 1.2, the peaks at the highest cooling rates will

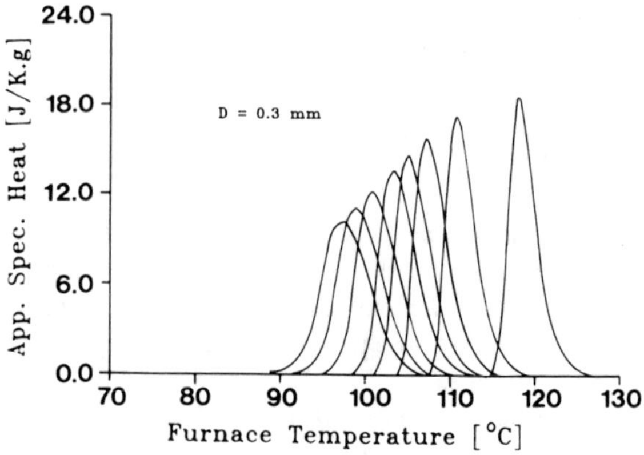

Fig. 1.5 Courses of apparent latent heats as functions of furnace temperature for the nominal cooling rates (from left to right) of 50, 40, 30, 20, 15, 10, 5 and 1 K/min and a sample of PP of a thickness of 0.3 mm, according to Wu's calculations [35]. Courtesy of Springer Verlag

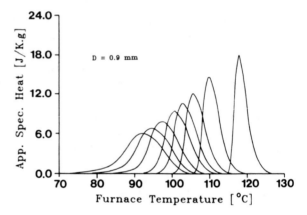

Fig. 1.6 Courses of apparent latent heats as in Fig. 1.5, but for a sample thickness of 0.9 mm [36]. Courtesy of Springer Verlag

become lower and broader than in Figs. 1.5 and 1.6. This means that the approximation of the first paragraph leads to an underestimate of the effect.

In this connection we have to admit that the heat transfer coefficient γ is lowered in our experiments a little bit by the fact that three knobs were punched in the bottom of the pan in order to improve the reproducibility of the measurements. In fact, the bottoms of the pans are never completely flat. But the reader should accept that even with an optimum touch of the pan the corresponding heat transfer coefficient will not rise to infinity, as some optimists seem to believe. But this means that the general picture provided by Figs. 1.5 and 1.6 will remain unchanged. And a somewhat decreased heat transfer coefficient will be of advantage for a simplified treatment of the problem, as will be shown right away.

At the end of this section it must be admitted that the influence of the nucleation, which is caused by the surfaces of the walls, is ignored. In fact, Schneider's equations are applied in the form, which only holds for an unlimited space. An extended study of the influence of the walls will be given in Sect. 1.4.2. From this study it appeared that the wall effect would form a prohibitive complication of Wu's calculations. In fact, they are involved enough. In Sect. 1.4.2.3, however, it will be shown that the corresponding error decreases with increasing cooling rate. This means that the difference between the graphs in Figs. 1.5 and 1.6 remains very realistic.

1.3.3.2 A Simplified Analytical Solution

As a first step a comparison is made with the results obtained for the asymmetric case. For the pertinent calculations the full value of γ was used on the contact side. On the other side a zero heat transfer coefficient was assumed. It appeared that for a thickness $D_s = 0.3$ mm the results for the asymmetric case were practically identical with the results depicted for the symmetric case in Fig. 1.5 for that thickness. In particular this held for the highest, most critical cooling rate of 50 K/min. So, practically no difference was found for the two limiting cases, where the reality lies in between. But this means that the results are dominated by the relatively high resistance against heat transfer between furnace and pan. In fact, the internal heat transfer problems seemed to play only a minor role. However, for the larger sample thickness of $D_s = 0.9$ mm the situation was quite different. For this larger thickness of the sample the peaks became considerably broader for the asymmetric case. Apparently, with the larger distances for heat flow within the sample it became of importance, whether heat flew to both sides or only to one side. But this meant that a criterion had to be found for the license to believe that it does not matter, where exactly on the sample surface the heat transfer takes place.

For the purpose, the authors of [35] defined kind of a Nusselt number N_u^* as the ratio of the outer heat transfer coefficient γ and the sample intern heat transfer coefficient γ^*, i.e.:

$$N_u^* = \frac{\gamma}{\gamma^*}, \tag{1.56}$$

where γ^* is defined as

$$\gamma^* = \frac{4\,S\,\lambda_s}{D_s}. \tag{1.56a}$$

The authors observed that for a diameter of the tablet of 3 mm, a thickness of 0.3 mm, an outer heat transfer coefficient of 16 mW/K and a heat conductivity of 0.193 W/mK the said Nusselt number was:

$$N_u^* = 0.11. \tag{1.56b}$$

For the thicker sample of 0.9 mm a value of 0.33 was found. The conclusion is that N_u^* must not be larger than one tenth, if a simplified method, which will be explained right away, can be used. In fact, N_u^* can always be calculated with the aid of the available data. It should be clear that for metallic samples N_u^* is always extremely small because of the large values of λ_s. (For indium one has $\lambda_s = 76.2$ W/ mK, which is about 400-fold the value given above for PP.)

In the simplified model one can content oneself with a simple balance equation, which reads:

$$-(m_s c_s + m_p c_p)\frac{dT_s}{dt}(t) + m_s h_s \frac{d\xi}{dt}(t) = \gamma(T_s(t) - T_f(t)) \tag{1.57}$$

with

$$T_s|_{t=0} = T_{rf}, \tag{1.57a}$$

where T_s is now "the" sample temperature. The "apparent latent heat" reads now:

$$l(t) = \frac{\gamma\,(T_s(t) - T_b(t))}{m_s q} \tag{1.58}$$

where the "base line" is

$$T_b(t) = T_f(t) + \frac{q}{\alpha}, \tag{1.59}$$

with T_f according to Eq. (1.53). In fact, the straight base line runs parallel to the straight line giving the furnace temperature, but is delayed in the present case of cooling by the second term on the right side of the equation (see Ref. [34]). For α see Eq. (1.61) below. This line shows the influence of the heat flow, which occurs without crystallization because of the cooling effect. It is evident that the heat capacities of sample and pan are not zero.

Using these equations one obtains DSC-curves which are almost indistinguishable from the curves shown in Figs. 1.5 and 1.6. For the thicker sample, for which N_u^* seems to be too large, this fact is a little surprising.

Now we arrive at the point, where the determination of the heat transfer coefficient becomes urgent. This determination is based on the fact that the DSC-signal returns to the base line according to an exponential function of time, as soon as the phase transition (crystallization) comes to an end. For the present purpose this moment is chosen as time $t = 0$. For this initial condition and for $d\xi/dt = 0$ the solution of Eq. (1.57) reads:

$$T_s(t) - T_b(t) = (T_s(0) - T_b(0))\,e^{-\alpha t} \tag{1.60}$$

with

$$\alpha = \frac{\gamma}{m_s c_s + m_p c_p}. \tag{1.61}$$

Aiming at a linear dependence on time the logarithm is taken on both sides of Eq. (1.60):

$$\ln\left(T_s(t) - T_b(t)\right) = \ln\left(T_s(0) - T_b(0)\right) - \alpha t. \tag{1.62}$$

At this point Eq. (1.53) must be reconsidered. Because of this linear relation one can consider T_f also as a time parameter. For the use of this parameter Eq. (1.62) must only be multiplied by q on both sides. If Eq. (1.53) is rearranged into $-qt = T_f(t) - T_{rf}$ one obtains the following equation, where the logarithm of the apparent latent heat is involved.

$$q\,\ln(l(T_f)) = f(q) + \alpha\,T_f \tag{1.63}$$

The recipe for the determination of α, to which the heat transfer coefficient γ is related by Eq. (1.61), is now obvious. The logarithm of the DSC-signal must be multiplied by the cooling rate q and plotted against the furnace temperature. The slope of the straight line, which is obtained, if the right temperature range is chosen, gives the value of α. In Fig. 1.7 this is demonstrated for an industrial PP.

The cooling rates are indicated near the curves. The expected linearity and parallelism is clearly shown. The slight curvatures at the lower ends must be ascribed to small errors in the base line position. The value of γ, which is obtained

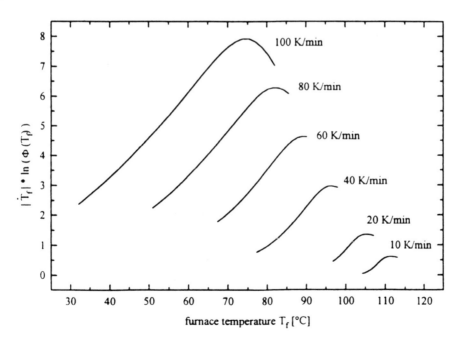

Fig. 1.7 Plot of the DSC-curves according to Eq. (1.63) for an industrial PP. The cooling rates are indicated near the curves [22]. Courtesy of VCH Wiley, Weinheim

with the aid of Eq. (1.63), is a little lower than the value of $\gamma = 16$ mW/K, as used in Wu's calculations, i.e. $\gamma = 10$ mW/K. This fact can easily be explained: In Ref. [34] metals were used for the samples. As the heat conductivities of these samples is factors of about 400 higher than that of the polymer, the obtained heat transfer coefficient is higher and practically equal to the coefficient for the transfer between pan and furnace. In the case of the polymer the effective temperature in the sample is an average temperature. As a consequence, the heat transfer coefficient is also an effective one, which is influenced by the low conductivity of the sample.

The balance Eq. (1.57) can be used for the derivation of an interesting equation, which relates the degree of crystallinity, which grows during the cooling cycle, to the development of the DSC-curve. For the purpose time t is replaced by T_f according to Eq. (1.53) and the integration is carried out with respect to the latter variable. For this procedure it is important to realize that the heat flow

$$\Phi(t) = \gamma \left(T_s(t) - T_f(t) \right) \tag{1.64}$$

which is found on the right side of Eq. (1.57), comprises not only the flow of the evolved latent heat $m_s h_s$ but also that of the stored heat $(m_s c_s + m_p c_p)$ of the system. So, one has

$$\Phi(t) = \Phi_{cr}(T_f(t)) + \Phi_{st} = \Phi_{cr}(T_f(t)) + q\,(m_s c_s + m_p c_p), \tag{1.65}$$

where subscript cr stands for crystallization and subscript st for stored heat, which is removed in creating the base line $T_b(T_f)$. The differential Eq. (1.57) is properly transformed into a differential equation with T_f as the independent variable. When this equation is integrated, Eq. (1.65) is substituted. In this way one arrives at:

$$\xi\left(T_f\right) = \frac{\Phi_{cr}(T_f)}{\alpha\, m_s h_s} + \frac{1}{m_s h_s q} \int\limits_{T_f}^{T_{rf}} \Phi_{cr}(u)\, du. \tag{1.66}$$

This equation was derived by G. Eder in the course of the preparation for our review of 1997 [22]. One may observe that the upper temperature T_{rf} lies outside the peak area of the DSC-measurement. As a consequence only the lower temperature T_f of the integral is of importance. However, if the cooling is continued to the lower end, the well established integral over the whole peak is left over in the second term of Eq. (1.66). In fact, Φ_{cr} is zero again at that lower end. From this integral one obtains the degree of crystallinity of the just solidified sample. If on the other side an intermediate temperature T_f is considered, one has to bear in mind also the first term. It is now ten years from the publication of this result. So far, as we can take in the situation, nobody has ever cared of this fact. This first term depends on the heat transfer coefficient through α (Eq. (1.61)), which means that this coefficient has to be determined anyway.

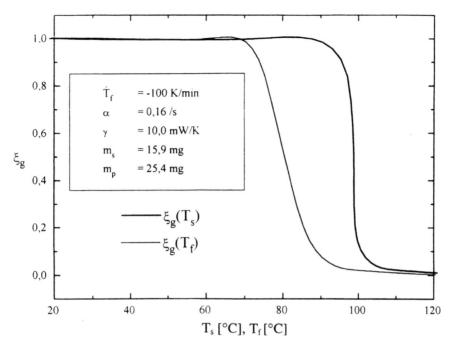

Fig. 1.8 The left curve gives the course of the crystallization of an industrial PP according to Eq. (1.66) for the high cooling rate of 100 K/min. For this curve the furnace temperature takes over the role of the time. The right curve gives the course of the crystallization as a function of the real sample temperature according to Eq. (1.67). See Ref. [22]. Courtesy of VCH-Wiley, Weinheim

Equation (1.66) enables the calculation of the progress of crystallization as a function of the time variable T_f. This is shown by the left curve of Fig. 1.8 for an industrial PP, which has also been used for the measurements leading to Fig. 1.7. For the purpose a high cooling rate of 100 K/min was applied. The furnace temperature T_f takes over the role of time in this curve.

The high cooling rate was achieved in a special DSC-apparatus at Philips Research in Eindhoven. This apparatus was cooled with liquid nitrogen. Nowadays there are already machines, which can be cooled much faster [33]. The present authors take the liberty to ask, whether the pertinent authors want to entangle themselves in insoluble heat transfer problems inherent to the architecture of the traditional machines.

The second curve in Fig. 1.8 invites us immediately for a discussion. It will turn out that with higher cooling rates a correction of the peak temperature becomes unavoidable, as also Eder has shown. As already announced, the temperature T_s in the sample becomes quite different from the furnace temperature T_f at the maximum of the DSC-curve. In fact, according to a combination of Eq. (1.59) with $\Phi_{cr} = \gamma$ $(T_s - T_b)$ one has:

$$T_s(t) = T_f(t) + \frac{q}{\alpha} + \frac{\Phi_{cr}(T_f(t))}{\gamma} \qquad (1.67)$$

The sum of the second term and the third term of this equation can give quite a temperature rise above the furnace temperature. This becomes evident for the high cooling rate of 100 K/min. In fact, with the aid of this equation the transformation from T_f to T_s was carried out on the abscissa of Fig. 1.8. In this way the right curve of this figure is obtained. It shows that with a nominal cooling rate of 100 K/min crystallization occurs with this polymer virtually at a constant internal temperature of about 99 °C. At this point the reader is reminded of the considerations in Sect. 1.3.2, where the dimensionless classification number is defined. In this connection it has been assumed that crystallization occurs at a definite temperature above the temperature, where the growth speed of spherulites shows its maximum. A look on Fig. 1.2 informs us that this maximum lies for PP at about 90 °C.

If after a change of the scale of the abscissa (with the use of Eq. (1.53)) the left curve of Fig. 1.8 is differentiated with respect to time, one has only to multiply the result by $m_s h_s$, in order to obtain the real rate of crystallization $m_s h_s d\xi/dt$. This rate is plotted in Fig. 1.9 (right curve) against the furnace temperature T_f. For comparison Φ_{cr} is also plotted against T_f (left curve). One notices the difference. But the

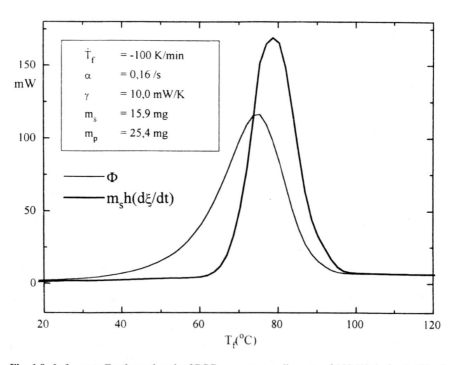

Fig. 1.9 Left curve: Exothermal peak of DSC-curve at a cooling rate of 100 K/min for the PP of the previous figures. Right curve: Heat evolved by the sample in the time span corresponding to the span of T_f. Both curves as functions of T_f. See Ref. [22]. Courtesy of VCH-Wiley, Weinheim

right curve is not a real DSC-curve: the internal rate of cooling is not constant because of the internal re-heating effect.

1.3.3.3   Scanning Micro-Calorimeter for High Cooling Rates

The two previous sections were not formulated with the intention to advocate them for practical use. In fact, from the onset it was clear to us that cooling rates, as occurring in processing praxis, are much higher than those, which can be realized in DSC-machines of traditional architecture. The goal of the just mentioned sections was a more general one. These sections were considered to be very useful for a demonstration of the difficulties, which arise in principle with scanning calorimetry and other measuring techniques helpful for the evaluation of processing parameters. Our group has never been engaged seriously in those techniques before. But we became aware of the central role, which those measurements play in the understanding of polymer processing. Another point was a theoretical one. In fact, the calculations by Mrs. Wu clearly demonstrate the usefulness of Schneider's rate equations, which cannot easily be replaced by other equations.

Recently, however, a research group at Rostock University has developed a micro-calorimeter, which deserves a serious consideration in the present context [37–41]. In fact, this calorimeter enables extremely high cooling rates. One has to consider this development as a break-through in scanning calorimetry. For the purpose samples of about 100 ng are placed on the middle of a thin silicon nitride membrane, which has a thickness of a few μm and a diameter of about 2 mm. Below the sample there is a tiny electric heater surrounded by the hot junctions of six thermopiles, of which the cold junctions are on the frame supporting the whole, which is immersed in a closed gas atmosphere. The nature of the gas (nitrogen, helium), its pressure and its temperature can be varied for the purpose of more or less severe cooling. Such an apparatus can be used for a lot more applications than those, which are of interest in the present study, where the consequences of rapid cooling attract our attention.

In the first paper mentioned [37] one finds as Fig. 1.4 a graph of particular interest. In this graph the apparent heat capacity (latent heat) of a 120 ng sample of a special polyethylene of a narrow molar mass distribution is plotted against the temperature of the heater for various extremely high cooling rates indicated. This apparent heat capacity is deduced from the power input of the heater, which is controlled for pertinent linear decreases of the heater temperature with time. As expected, the locations of the peak maxima shift to lower temperatures with increasing cooling rates. The big sensations are the high cooling rates. The location of the peak, which is obtained at a cooling rate of only 10 K/min with an ordinary DSC-machine, is indicated at the abscissa by an arrow.

The fact that the peak for a cooling rate of 140 K/s is lower than that for 450 K/s, cannot easily be explained. From kinetic data of HDPE Eder [28] estimated a minimum cooling rate of at least 10000 K/s (Table 1.1 in Sect. 1.3.1) for bypassing crystallization. This result seems to fit nicely into the picture delivered by Fig. 1.10.

The fact that the peaks become broader and lower with decreasing temperature of the maximum, can have various reasons. There is the heat transfer problem which, however, cannot easily be solved because of the fact that the sample experiences two heat flows, one from the heater and the other to the gas of constant low temperature. But there is also a possible influence of changing crystallization kinetics with decreasing temperature. Mrs. Wu assumed for her calculations an undisturbed linear increase of the logarithm of the growth speed with decreasing temperature (tangent on the half logarithmic plot at higher temperatures). If sample and furnace have equal temperatures (no heat transfer problems), this assumption leads to equal peak heights [36]. However, with temperatures lower than 400 K one finds with HDPE a considerable deviation to lower values from the said tangent of the logarithmic growth speed. The temperature of the maximum of the growth speed of spherulites of HDPE must lie at about 300 K (half way between melting point and glass transition). Our measurements [9] (see also [22, 42]), which reach down to about 350 K, already indicate the approach of that maximum. So, the flattening of the peak in Fig. 1.10 can also be explained in this way.

Another very interesting result of the Rostock apparatus is shown in the fifth publication cited above (De Santis et al. [41], Figs. 1.5 and 1.6). In these figures the behavior of a sample of iPP is presented. The first of these figures is reproduced here as Fig. 1.11.

At the lower cooling rates, between 30 and 120 K/s, one observes peaks, which can be ascribed to the α-modification. This fact was verified by the authors in heating experiments at 50 K/s. The typical melting peak at 160 °C was found. With further increasing cooling rates one observes the upcoming of a peak, for which the

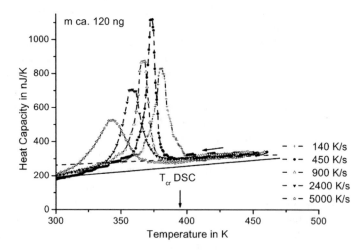

Fig. 1.10 Temperature dependence of the apparent heat capacity of a sample of a linear polyethylene of about 120 ng for various cooling rates according to Ref. [37]. (From left to right 5000, 2400, 900, 450 and 140 K/s.) Dashed and solid straight lines: temperature dependence of heat capacities of permanently amorphous and crystalline samples of corresponding mass. Courtesy of Elsevier

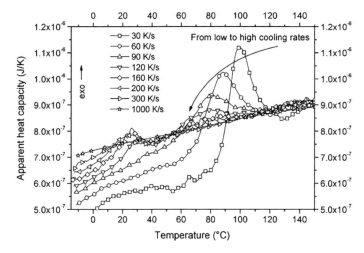

Fig. 1.11 Thermograms of a sample of iPP for various selected rates of 30, 60, 90, 120, 160, 200, 300 and 1000 K/s according to Ref. [41]. Courtesy of American Chemical Society

mesophase is responsible. With a cooling rate as high as 1000 K/s no crystallization at all could be registered. The obtained straight line depicts the decrease of the heat capacity of the amorphous material with decreasing temperature. Interestingly, after the formation of the α-peak the heat capacity submerges with farther decreasing temperature below the value of the amorphous material by about 30%. This dip seems to reflect the lower heat capacity of the crystalline material. For the first time this figure gives a complete analysis of the processes occurring during cooling of a sample of iPP. Also the mesophase could be identified by the typical thermogram, which was found with re-heating.

Again, there seems an interesting comparison with the results of Table 1.1 of the present publication. In this table one finds for iPP a minimum cooling speed for bypassing crystallization of 60 K/s. From Fig. 1.11 one learns that 160 K/s were needed for suppressing the α-peak. It appears that this value is higher than the value calculated by Eder from the kinetic data. However, we find this fact satisfying: Our results, as depicted in Figs. 1.1 and 1.2 of the present publication, seemed too exaggerated to some critics. However, in order to make the result of Eder´s calculation agree with the above result of a thermal measurement, at least higher maximum values of the growth speed will have to be assumed.

For those interested in the heat transfer queries, Fig. 1.6 of the said paper seems to be informative. This figure is reproduced here as Fig. 1.12.

The authors of the paper think that the lower positions for the standard DSC are a consequence of a relatively smaller surface. An alternative idea, however, is that the positions from DSC are too low because of insufficient heat transfer. In this connection Fig. 1.9 should be considered. If this is true, one can conclude that the heat transfer problem is less urgent with the nano-calorimeter.

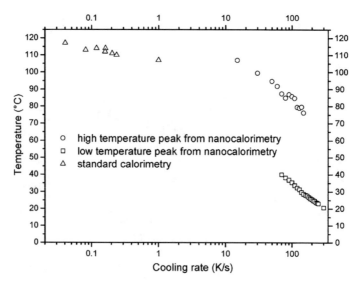

Fig. 1.12 Exothermal peak positions versus cooling rates, as obtained for an iPP [41]. Circles: α-modification, squares: mesomorphic phase, both according to nano-calorimetry, triangles: α-modification from standard DSC. Courtesy American Chemical Society

1.3.4   Phenomena of Propagation and Spreading

It will be clear to the reader that a situation, where crystallization starts simultaneously as a homogeneous process in a large volume, is a very unrealistic situation. Theoretically, such a process can happen only, if the temperature is everywhere the same. But such a homogeneous temperature can be obtained only after a thermal equilibration at a certain degree of undercooling, and the material is not permitted to start crystallization within the time span, which is needed for this equilibration. At the onset of every solidification process the bulk of the molten material is still at a temperature above the melting point. But this means that crystallization starts in reality always at some boundary, where the cooling process sets in. A particularly drastic process is initiated by a quench of the wall of the vessel, in which the melt is contained. But every quench takes some time. Even if a thin metal wall of one to two millimeters thickness is quenched on its outside with the aid of a streaming heat transfer fluid, it can take some seconds until its inner surface has reached the desired constant temperature. The present section is devoted to all kinds of problems arising around these queries.

1.3.4.1   Fast Crystallizing Polymers

The propagation of the crystallization in HDPE, starting at a quenched wall, has been measured by Ewa Ratajski [7]. For the purpose two versions of an apparatus were developed. In both cases the polymer was inserted as a solid cylindrical piece of a diameter of 20 mm and a height of 10 mm. With one apparatus the cylindrical piece was cut into two pieces under a certain angle with the axis. Before the obtained two pieces were joined again, a strip of black paper was interposed. The sample, as prepared in this way, was shoved into a tube of glass or aramide. This tube could be heated with the aid of a stream of hot air and indirectly by heat conduction from metal tubes situated on both ends. On one side the glass tube was closed by a metal plate, which could be quenched with the aid of a heat transfer fluid. On the other side there was a glass plate, through which the sample could be inspected. The whole arrangement was heated to a temperature, where the polymer sample melted. After equilibration the metal plate was quenched to a previously chosen temperature and the hot air was adjusted to a temperature between the original temperature and the temperature of the heat transfer fluid. As a consequence crystallization started at the quenched wall and propagated in axial direction into the sample. On the black strip this propagation could be followed because of its inclined position with the aid of a cathetometer. An inspection from aside is impossible because of the strong temperature gradients in the sample, by which the light beam is deflected into the direction of the lower temperature, an effect which causes an overestimate of the thickness of the crystallized zone. In the second version of the apparatus an inspection from aside was permitted. But a microscale printed on a strip of white paper was inserted into the heart of the sample in axial direction or at an angle of 30° with the axis (for a better illumination only).

Results of measurements, which were obtained by E. Ratajski for a sample of HDPE, are reproduced in Fig. 1.13. With HDPE the transition from the transparent melt to the opaque solid is quite sharp. Nevertheless, this transition occurs in a diffuse crystallization zone, which is relatively narrow for HDPE. The contrast with a real crystallization front will be discussed below.

It turns out that the agreement between the elder and the newer measurements is excellent (see the caption to the figure). One notices that the curves do not go through the origin. However, at the lowest applied wall temperature of 100 °C it becomes difficult to recognize this fact. The shape of this curve becomes quite similar to curves, which are obtained with the classical square root law (see below). In an early paper Berger and Schneider [43] have shown that one can indeed expect curves of a shape similar to the curve obtained with the square root law. Only, these curves are shifted down parallel to the ordinate axis. The higher the wall temperature the larger the necessary shift. In this way they cut the time axis at finite values indicating the room for transcrystallization.

If a polymer is investigated, which shows a much lower crystallization speed like iPP, one obtains curves, for which the distances in time between the origin of the graph and the apparent intersections with the time axis are larger. Of course, for the purpose the said curves, as drawn through the experimental points, have to be continued down to the abscissa. For instance, this effect is characteristic for an industrial PP, for which Fig. 1.14 is prepared. But for such a polymer also the transition from the clear melt to the solid is more gentle. The crystallization zone is broader and more diffuse and cannot be determined as accurately as with HDPE. On

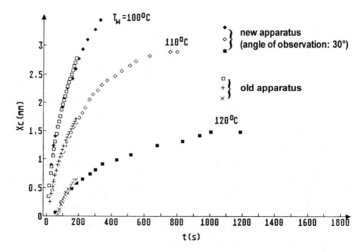

Fig. 1.13 Thickness of the crystallized layers as functions of time for a series of temperatures of the quenched wall for a sample of HDPE. Initial homogeneous melt temperature T_i = 170 °C, wall temperatures 120, 110, 100 °C counter-clockwise, according to E. Ratajski [7]. Stars, diamonds and closed squares were obtained with the newer apparatus, open squares and two types of crosses are from to the elder apparatus

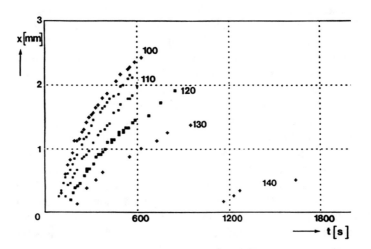

Fig. 1.14 Thickness of the crystallized layers as functions of time for various temperatures of the quenched wall for an industrial PP. Initial homogeneous temperature of the melt T_i = 200 °C, wall temperatures 140, 130. 120, 110, 100 °C counter-clockwise, according to E. Ratajski [7]

the other hand one can clearly see that at short times there must be a supplementation for the necessary connection with the origin. In fact, the curves have to start necessarily with a finite slope at time zero. (By the way, this is also the crux with the square root law, which starts with an infinite slope. No real process can start in this way. For a proper discussion see below.)

In fact, if the temperature of the quenched wall is not too low, and the said distance to the origin is still large enough, one finds the so-called "transcrystallization" within the corresponding time span. This term has been introduced by Wunderlich [23] some time ago. This name has to do with the fact that with a very thin sample this type of crystallization propagates without hindrance from one to the other surface of the sample, if a proper, not too low temperature is chosen. In our case this propagation is hindered after some time by the diffuse crystallization in the bulk of the sample.

The transcrystallization has its origin at nuclei located in the wall surface, whereas the diffuse crystallization is initiated by nuclei located in the melt itself. If the influence of the released latent heat is to be taken into account, a correct theoretical description of this situation can be carried out only with the aid of Schneider's rate equations [26] and the kinetic data presented in Sect. 1.1. Ratajski and Eder [7] assumed that the influence of the latent heat is not too strong. So, the time dependent spatial temperature distribution (error function solution) was used for an amorphous fluid and the crystallization kinetics were engrafted on this distribution by local integration over the time. The result was quite realistic. Some justification will be tried in the next section. However, for a complete description, including the upcoming hindrance by the crystallization in the bulk, only the complete theory for samples of restricted size will be helpful. A later section will be devoted to this difficult subject.

As an illustration a photograph is taken over, on which the effect of transcrystallization is shown for an industrial PP. In this case the wall was rubbed in with Cinquasia Gold as an agent for β-nucleation. As a consequence of this preparation of the wall the transcrystalline layer contains only the β-modification. Such a layer grows faster than a layer of the α-modification and gives a more pronounced effect. By the way, in Wu's calculations, aiming at a presentation of the heat transfer effect in differential scanning calorimetry, any influence of transcrystallization was disregarded. However, for the lower peak temperatures, which are reached with higher cooling rates, this simplification cannot be of a larger thermal influence. In fact, those transcrystallized layers could be observed in cross-sections taken from samples, which were removed from the pan of the DSC-apparatus. The higher the cooling speed was the thinner were the transcrystallized layers [22] (Fig. 1.15).

Interestingly, within the transcrystallized zone one finds some enclosed spherulites of the α-modification. Apparently, these spherulites were nucleated at temperatures higher than the temperature of the quenched wall, where only a few α-nuclei were present (see Fig. 1.1). (In fact, the melt itself was not nucleated for the β-modification.) During the growth of the said α-spherulites the crystallization front of the β-modification arrived from the wall. Nuclei of the α-type of a larger number, as characteristic for the lowered temperature, were certainly covered by the front growth, before they could reach a noticeable size. There are still two remarks to be made: The birefringence of the β-crystals is much higher than that of the α-

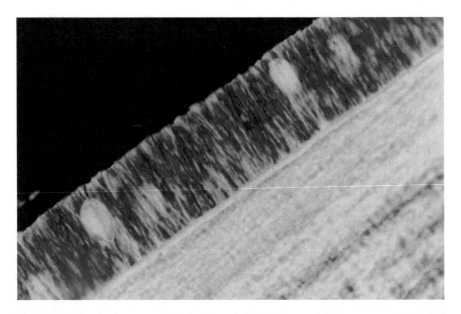

Fig. 1.15 Undisturbed transcrystallization for an industrial PP at an initial temperature in the bulk of 200 °C, a quenched wall temperature of 110 °C and a contact time of 49 s, if the β-modification is initiated at the wall [7]

crystals, because of epitaxial growth of the latter crystals. This was pointed out by Turner-Jones et al. [44]. This fact may also be the reason for the slower growth.

The enclosed α-spherulites have the shape of pears. One can see that their growth was hampered by the β-growth, particularly on the apex turned towards the wall. On the tail there was a running match, which was finally won by the faster β-growth. These phenomena were discussed first by Lovinger et al. [45].

1.3.4.2 Theory of the Front Growth

The theory of front growth is quite instructive, even if front growth is always restricted with polymers to a short time span, as the previous section has demonstrated. In fact, the zone moves faster comparable with a well performed relay race, where with every handling over of the baton one arm length (spherulite radius) is won. The classical theory for the front growth yields the famous square root law. As a first step one needs the heat balance at the moving boundary. For the progress of crystallization starting at a cold wall one has:

$$\lambda_c \left(\frac{\partial T}{\partial x}\right)_c = \rho\, h\, \frac{dx_c}{dt} + \lambda_l \left(\frac{\partial T}{\partial x}\right)_l . \tag{1.68}$$

In this equation subscripts stand for liquid ("*l*") and crystalline ("*c*"). For the heat conductivities λ is used, h is the latent heat, taken positive for the crystallization, ρ is the density, averaged over the two phases, x is the distance from the wall, T is the temperature and t is the time. The term on the left side gives the amount of heat removed by conduction from the moving boundary through the solid, the first term on the right gives the latent heat accompanied with the speed of the moving boundary, and the second term on the right gives the amount of heat conducted into the boundary from the still hotter melt. This equation must always be fulfilled.

However, Neumann [46] assumed that the temperature at the boundary is constant and equal to the temperature T_m at the equilibrium thermodynamic melting point. In this way it was easy to fit the two well-known similarity solutions of the equation of heat conduction.

$$T_w + A \, \text{erf} \left(\frac{x_c}{2\sqrt{a_c t}} \right) = T_\infty + B \, \text{erfc} \left(\frac{x_c}{2\sqrt{a_l t}} \right) = \text{const.} = T_m \qquad (1.69)$$

In this equation A and B are still undetermined constants, and a_l and a_c are heat diffusivities of melt and solid. Equation (1.69) consists of two equations. The function erfc(u) is the complementary error function $(1 - \text{erf}(u))$. Because Eq. (1.69) must be valid for all values of time, x_c must obey the following equation:

$$x_c(t) = 2\kappa \sqrt{a_c t} \qquad (1.70)$$

As Eqs. (1.69) and (1.70) are three equations for the three unknown factors A, B and κ, one can now calculate the values of these factors. Of particular interest is factor κ showing up in Eq. (1.70). However, the analytical solution is quite complicated [46] and will not be reproduced here. The only important fact is that κ decreases with increasing latent heat. The more heat is released in the boundary, the lower is the speed, with which the boundary can move.

As already mentioned, this square root law cannot be valid for short times. There is no real process, which can start with infinite speed. As a consequence, one has to look for a better solution. In fact, the condition of a constant temperature T_m at the moving boundary means that crystallization must occur without delay, when this temperature is reached during the cooling process. In reality, however, one has always some degree of undercooling, before the process can start with finite speed.

In the modified model, as developed by Eder and Janeschitz-Kriegl [47] (see also [36]), the temperature T_c at the moving boundary is no longer invariant. It becomes part of the solution. The speed of the movement of the boundary is determined by this temperature. The assumption is made that this speed is equal to the growth speed of spherulites at that temperature. Remarkably, at T_m this speed is equal to zero. As the growth speed of spherulites can be determined experimentally (see Fig. 1.2), also dx_c/dt can be considered as a known function of T. Eder calculated the temperature T_c at the boundary for a sandwich confined between two parallel plates. Before the experiment is started, both plates are at T_i above the equilibrium melting point. At time $t = 0$ one of the plates is quenched to temperature T_w below the melting point. No calculations are known for the semi-infinite space. However, at low Fourier numbers $Fo = ta_l/D^2$, where D is the thickness of the slab, there will be no noticeable difference between the result for the slab and for the semi-infinite space. An iterative numerical procedure was necessary, because of the fact that for every chosen T_c the influence of the corresponding dx_c/dt had to be checked, until Eq. (1.68) was satisfied.

The following data where used for the calculation: $T_i = 245$ °C, $T_w = 90$ °C, $T_m = 160$ °C, $T_g = 60$ °C, $a_l/a_c = 0.54$, $\lambda_l/\lambda_c = 0.60$ and $D = 1$ mm. A parabolic temperature dependence of the growth speed as a function of temperature was assumed, which went through zero at T_m and at the glass transition temperature T_g. At its maximum, halfway between T_g and T_m, a maximum growth speed of 5.2 mm/ s was assumed. Later it was realized that this maximum growth speed was chosen much too high. For the fastest crystallizing polymers, namely HDPE or polyketones (see Table 1.1) this maximum growth speed was found to be at least a decade lower. However, the calculations keep their value because not only the speed of the layer propagation but also the development of the diffuse crystallization, which finally hampers the propagation of the growth front, obeys the same temperature dependence, which has been exaggerated for the said calculation.

Dimensionless parameters have been used for the presentation of the results, namely $x^* = x/D$, $t^* (= Fo) = ta_l/D^2$, $\Lambda (=St) = h/c_l(T_i - T_w)$ and $\Theta_c = (T_c - T_w)/ (T_i - T_w)$. From Fig. 1.16 one learns that all values of x^* start with a linear part at low values of the Fourier number t^*. At large Fourier numbers the value of x^* is approached, which holds for the stationary state, as obtained between the hot and the cold walls of constant temperatures. The Stefan number Λ is of great influence only at high values. With polymers one can assume that the Stefan number is always of the order of one. Also the result of the square root law is indicated in Fig. 1.16 by the not very accurate course of $(at)^{1/2}$. This law, which is drawn up for the semi-infinite space, certainly holds as long as x_c^* remains small compared with unity. At this point it must be emphasized, however, that with the use of a realistic growth speed of only one tenth (see above) the discrepancy with $(at)^{1/2}$ will be many times larger.

In Fig. 1.17 the reduced temperature Θ_c, as obtained at the moving boundary, is shown as a function of the Fourier number t^*. For relevant Stefan numbers Λ this temperature remains close to the wall temperature for small values of the Fourier number. According to the classical theory Θ_c should be constant and equal to Θ_m. The results are according to Ref. [47], courtesy of Springer-Verlag.

In conclusion one can say that at short contact times, where the diffuse crystallization does not yet hinder the front growth, the speed of this front growth is determined by the growth speed of spherulites at the temperature of the quenched wall. The prerequisite is that the average distance between nuclei on the surface of the wall is much smaller than the average distance between the nuclei in the melt. In fact, with a dense population of nuclei on the wall the spherical growth fronts, which are formed in the beginning around each nucleus, quickly unite to a flat

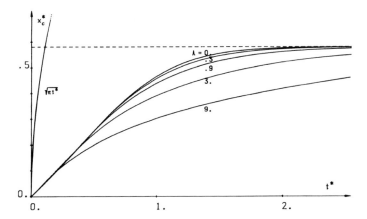

Fig. 1.16 Reduced distance x_c^* of the crystallization front from the quenched wall against the Fourier number t^* for various values of the Stefan number Λ for a slab of finite thickness according to Ref. [47]. Courtesy of Springer-Verlag

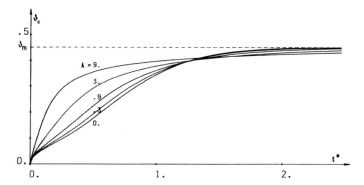

Fig. 1.17 Reduced temperature Θ_c at the crystallization front against the Fourier number t^* for various Stefan numbers Λ. According to the classical treatment for all Fourier numbers Θ_c is at the constant value of Θ_m

growth front, which moves into the melt. This fact will be used for the determination of growth speeds in a later section of this monograph.

1.3.4.3   Correctly Quenching a Polymer at a Metal Wall [48]

So far it was always assumed in Sect. 1.3.4 that a quench of the wall could occur fast enough, so that the wall reached the desired temperature within a time span short compared with the characteristic time of the crystallization process at that temperature. However, for fast crystallizing polymers this is a questionable assumption. In coping with this difficulty the hot polymer sample was suddenly brought into contact with a preheated metal wall of established temperature. This method was used in some critical experiments (see Ref. [7]). In this way the temperature dependence of growth speeds could be determined by an optical method, which will be described in a later section. (Light scattering arising at the surface.) However, with extremely fast crystallizing polymers like HDPE an improvement of this method appeared necessary nevertheless for temperatures below 90 °C. One has to look at Fig. 2.15, as given in Sect. 2.1.2, for the growth speeds of spherulites. In this figure a kink is found at 90 °C, where the full squares (for HDPE) abruptly fall off with further decreasing temperature. But such an abrupt deviation from a curve, which should be smooth, cannot be explained easily. The suspicion arose that there could be a heat transfer problem (re-heating by latent heat).

For an elucidation of this problem a machine was used, which originally was designed as a dilatometer. This dilatometer should permit an extremely fast quench of a thin slab of the polymer, so that a further decrease of the volume could be assigned to the crystallization kinetics. However, with preliminary temperature measurements in one of the metal walls we learnt that our goal could not be reached along this path. We called our machine a negative baking oven. Not the cold dough was pushed into the hot oven, but a thin layer of the hot polymer sample, as preheated on an aluminum foil in a special oven, was quickly shoved between two relatively cold parallel metal plates and sandwiched between these plates at time zero. The temperature at the newly formed contact surfaces between polymer and metal wall obeyed the following well-known equation:

$$\frac{T_{0,melt} - T_{interface}}{T_{interface} - T_{0,wall}} = \frac{\sqrt{(\rho.c.\lambda)_{wall}}}{\sqrt{(\rho.c.\lambda)_{melt}}} \tag{1.71}$$

In this equation ρ, c and λ stand for density, specific heat capacity and heat conductivity. The subscripts indicate the respective structure elements. As in particular density and heat conductivity of the metal are much larger than those of the polymer, only a slight increase of $T_{interface}$ above $T_{0,wall}$ was expected. For convenience, the relevant data for polyethylene, steel and copper are given in Table 1.3.

For the combination of HDPE and steel one obtains a temperature increase above the original wall temperature of 3.9 K, if the following temperatures are chosen for the materials in their state before the contact. For the wall of the machine: 40.5 °C (313.7 K), for the original temperature of the polymer: 150 °C (423.2 K).

Table 1.3 Physical Properties of relevant materials. For polymers see Ref. [49], for metals see Ref. [50]	Material	λ [W/mK]	ρ [kg/m^2]	c_p [J/kgK]
	Polyethylene	0.17	770	2100
	Steel	50	7900	500
	Copper	390	8900	390

However, for a sample thickness of 1.7 mm a quite curious course of the temperature was measured in one of the metal walls at a distance as close as possible to the contact surface. This course is shown in Fig. 1.18.

The whole procedure was repeated three times. A quite good reproducibility was obtained. During about three seconds the temperature close to the contact surface rose by almost ten degrees. As the temperature profile in the wall must show a continuous decrease from the contact surface to the outer surface, one can judge that the temperature at the contact surface arose by more than ten degrees. If a thinner polymer sample was used, the temperature rise was a little lower and the return to the temperature of the cooling water was a little earlier. If a measurement would be possible in the contact surface itself, one would find that in this surface the said temperature difference would start not at 0 °C but at 3.9 °C. This was confirmed by a numerical simulation. The wish-dream would be that the temperature difference in the contact surface would stay at those 3.9 °C instead of going up and down in an interval of ten degrees. But ten degrees is quite a lot for such a fast crystallizing polymer as HDPE. This fact becomes obvious also from a look at Fig. 1.13.

Interestingly enough, it is possible to achieve such a constant surface temperature, but not with a water cooled relatively thin steel wall [48]. As one of the authors (G.E.) has observed, the contact between two semi-infinite media is instructive in this respect. If the x-axis starts at the plane of the contact surface and both media are infinitely extended, one has the well-known solution of the equation of heat conduction in terms of the error function. With a constant interface temperature $T_{interface}$ such a solution reads for the metal on the positive side:

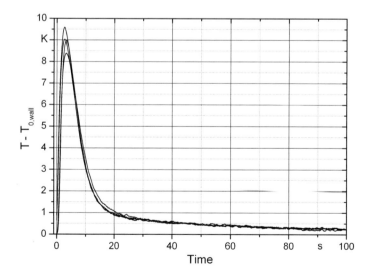

Fig. 1.18 Water cooled steel wall of 2 mm against sample of HDPE of 1.7 mm. Course of temperature versus time at a distance of about 0.4 mm from the contact surface. Original temperatures of wall and polymer were: 40.5 and 150 °C [48]. Courtesy of Hanser-Verlag

$$T(x,t) = T_{interface} - (T_{interface} - T_{0,wall})\operatorname{erf}\left(\frac{x}{2\sqrt{a_{wall}t}}\right) \qquad (1.72)$$

For the polymer on the negative side one has under this condition:

$$T(x,t) = T_{interface} + (T_{0,melt} - T_{interface})\operatorname{erf}\left(\frac{-x}{2\sqrt{a_{melt}t}}\right) \qquad (1.73)$$

If the derivatives with respect to x are formed of both equations for $x = 0$ and multiplied by the corresponding values of the heat conductivities, one obtains the heat flows at $x = 0$. These heat flows must be equal. In this way one obtains Eq. (1.71). As the time dependent terms cancel on both sides, the correctness of the assumption of a constant interface temperature, as calculated by Eq. (1.71), is proved for semi-infinite media. Admittedly. however, this proof only holds, if no crystallization occurs. The latent heat evolved would interfere with the chosen solution.

There is also another point: None of the media extends to infinity. As a consequence the condition $T_{interface} = $ const. only holds until the cooling effect (in the melt) reaches the mid-plain of the sample slab or the warming effect (in the wall) reaches the outer surfaces of the confinement. The events happening at such a moment will be discussed below.

For us it is of importance that one can imagine also a situation, where with finite distances to outer surfaces the constant contact temperature can be retained—at least, if no crystallization occurs. For the purpose the following train of thoughts is followed. A periodic sandwich construction, containing polymer and metal slabs alternatively, is envisaged. The corresponding values for the thickness of the layers must properly be adjusted. In order to keep the heat content in the polymer layer continuously equal to the heat content in—say—the slab of steel, one must find the right proportion. This proportion is given by:

$$D_{metal} = D_{polym}\sqrt{\frac{a_{metal}}{a_{polym}}} \qquad (1.74)$$

In this way one obtains the well-known periodic solutions for the cooling of a slab (of the polymer) or the heating of a slab (of the metal) at constant equal surface temperatures (see apposite text books). If one now imagines only a single polymer layer enclosed between two metal walls, one must assume that these walls are thermally insulated at their outside (with zero temperature gradient) and possess each one half of the thickness calculated with Eq. (1.74).With a thickness of the polymer layer of 1.7 mm one needs a thickness of 9.33 mm for two confining steel walls, which are insulated at their outer surfaces.

The next question is: What happens if the wall thickness is inadequate, in particular, if it is too small, and if cooling occurs by streaming water at the outside. There is a well-known equation for the penetration depth, if a temperature change occurs at the surface. This penetration depth means the depth, where the said change at the surface is felt for the first time. One has:

$$x_{pen}(t) = \sqrt{\pi a_{wall} t} \tag{1.75}$$

If the wall thickness is 2 mm, as in our case, the time, at which the penetration depth reaches this value, is 0.1 s. Looking at Fig. 1.18 one observes that the temperature maximum is reached at a much later time. With Eq. (1.72) one obtains for the difference $(T(x,t) - T_{0,wall})$ at the point (2 mm, 0.1 s) a value of 0.8 °C. If, however, at this distance of 2 mm there is no discontinuity in the metal wall, it makes sense to look at the differential quotient of Eq. (1.72) for the point (2 mm, 0.1 s). After a multiplication of this differential coefficient by the heat conductivity of steel one obtains a heat flux of 44.800 W/m². This should be the heat, which would be taken up per unit of time by a steel mantle much thicker than 2 mm. For a comparison the heat flux is estimated, which can be taken up by a fast stream of water at that distance of 2 mm. A heat transfer coefficient of 3.000 W/m²K has been reported [51]. With $\Delta T = 0.8$ K one obtains with this heat transfer coefficient only a heat flux of 2.400 W/m². But this means that with this configuration most of the heat, which can be conducted through a steel wall of a larger thickness, can no longer be drained off. But this means that there must be a temperature rise in the wall, as shown in Fig. 1.18. Actually, such a temperature rise can be found in a numerical simulation, as will be shown in the next paragraph. Such a numerical simulation will be unavoidable, if latent heat is evolved during a crystallization process.

The just described situation has been simulated for the initial temperatures for wall and polymer melt, which are given above. No latent heat was incorporated in the calculations. The pertinent results are shown in Fig. 1.19 [48]. In this figure the time axis stretches over only one second, in contrast to Fig. 1.18, where the time axis extends over 100 s. As a consequence the curves look much flatter.

First of all it must be mentioned that all curves go through a maximum at about two seconds, practically conform with Fig. 1.18. The pertinent figure is omitted here in saving space. The height of this maximum is at about 321 K for the uppermost curves, which corresponds with an increase of 7 °C above the initial wall temperature. The bigger height in Fig. 1.18 can be a consequence of latent heat produced in the real sample of HDPE. The extended time scale is chosen in order to show the initial horizontal part of the curve for the temperature in the contact surface. This horizontal part ends at about 0.1 s, in accordance with the simple result deduced from Eq. (1.75). At this point a temperature increase above the original wall temperature of 1.5 °C is obtained in these calculations. This increase is of the same order as the estimated increase of 0.8 °C, as mentioned above. If the thicknesses of the steel wall and the polymer sample are extended from 2 and 1.7 mm to infinity, the level of the short horizontal part is retained up to infinite times and the curves for increasing distances from the contact surface, which change for times longer than 0.1 s, snuggle adequately (see the original paper).

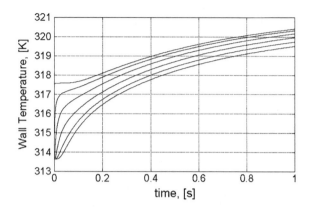

Fig. 1.19 Simulated courses of temperature in the steel wall for increasing distances from the contact surface (0, 0.23, 0.55, 0.97, 1.48 and 2.00 mm). The initial temperatures of melt and wall were 150 and 40.5 °C [48], conform with Fig. 1.18. No latent heat was taken into account. The curve for the temperature, which holds directly for the contact surface, is included as the uppermost curve. It starts at 317.6 K, as calculated according to Eq. 1.71, and shows a horizontal initial part, which ends at about 0.1 s. For a discussion see the text. Courtesy of Hanser-Verlag

An important point, of course, is the influence of the latent heat. As is obvious from the just presented results, a thin walled steel wall can be excluded from a further consideration. If a massive wall is preferred, the discussion remains about the most proper material. A numerical simulation was carried out for steel and for copper. HDPE was chosen as the polymer. A layer thickness of 1.7 mm was chosen as before. The thermal data of this polymer are contained in Table 1.3 of this section. For the calculation it is assumed that this polymer crystallizes quickly at 127 °C. The melting point is at about 145 °C. But crystallization kinetics are extremely sluggish at temperature close to 145 °C. Even at 127 °C there is no abrupt change to infinite speed of crystallization. So, one will still get an overestimate of the influence of crystallization. In Fig. 1.20 the result of the pertinent calculation is given for a copper plate of 63.6 mm thickness, being insulated at the outer surface.

Fig. 1.20 Courses of the temperatures in an externally insulated copper wall of 63.6 mm thickness at increasing distances from the contact surface of 0, 1.0, 4.0, 10.0, 35.0 and 63.6 mm. Initial temperatures for a 1.7 mm layer of HDPE and for the massive copper wall are 150 and 40.5 °C

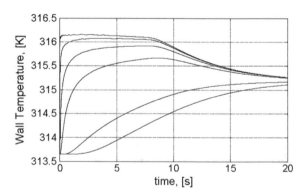

In absorbing also the latent heat of the HDPE layer of 1.7 mm the thickness of an externally insulated copper wall must be 63.6 mm, in contrast to steel, where this thickness must be 21.6 mm. For copper the contact temperature according to eq. (1.71) is 315.2 K, in contrast to steel, where this temperature is 317.6 K. From Fig. 1.20 one learns that for copper the crystallization process enhances the contact temperature very quickly to 316.2 K. From the accurate numerical data an increase of 0.95 K is obtained. For a steel wall a corresponding increase by 2.4 K is found, which means that copper is preferable. After about 7.5 s the plateau of Fig. 1.20 falls of. Finally the level of 315.2 K is reached, which holds according to Eq. (1.71) in the absence of latent heat. From the calculated course of the crystallization inside the sample of HDPE one learns that at this time of 7.5 s the crystallization is finished in the heart of the sample. However, as close to the surface crystallization can occur only earlier, the interesting processes will happen at times shorter than 7.5 s. So one can use the plateau temperature for the interpretation of data like those for transcrystallization.

Remains only the interpretation of the practical use of these informations. If one assumes that within these 7.5 s the interesting measurements can be done, one asks Eq. 1.75, how far the penetration has proceeded in this time. One obtains 18.3 mm. This is short compared with 63.3 mm. But this means that a thickness of 20 mm will be enough and no insulation will be necessary at the outer surface. However, laterally infinitely extended plates are also not operable. At best one can think of cylindrical peaces of a diameter of 20 mm and a length of 20 mm. Such cylinders can be insulated at their mantle to make them behave more like infinite plates. The biggest advantage is that one can insert thermocouples into the body of one of these cylinders. The required holes can be drilled into the body in a direction parallel with the axis, ending at various distances from the contact surface. As the couples have a lower conductivity than the body, the readings are not influenced by the conduction through the electrical leads, as is the case, if those couples are inserted into a cylinder of the low heat conductivity of a polymer. As already mentioned, the temperature profiles parallel to the axis are always showing a monotonic increase towards the contact surface. This means that an extrapolation to the surface will not be difficult and can be realized by a simple computer program, so that one can check the contact temperature every moment during the measurement. Unfortunately, such measurements have not yet been carried out. As a consequence, the present Sect. 1.3.4.3 can serve for the moment only as an enrichment of our insights.

1.4 Crystallization in Confined Spaces

So far this text has ignored or silently accepted the fact that every volume of a material is necessarily enclosed in a vessel or, at least, confined on some sides by walls. But walls are not ineffective with respect to the crystallization kinetics. In general they furnish extra nucleation centers. For instance, it has been assumed in

Sect. 1.3.4.2 without much ado, that the wall can nucleate a growth front, which moves into the sample. In contrast, with the simulation of the processes occurring in differential scanning calorimetry (see Sect. 1.3.3.1) any wall effect has been ignored. So it will be clear to the reader that the corresponding backgrounds still have to be elucidated and the errors evaluated.

1.4.1 A Derivation of Kolmogoroff's Equation According to G. Eder, Enabling a Generalization

In a fundamental paper [52] G. Eder was able to describe a promising route for this endeavor. The reader should also consult Ref. [22] for a somewhat more elaborate version. The author started with an equation taken from the theory of point processes [53]. For an illustration the author presented three point configurations. These configurations are shown in Fig. 1.21. for two dimensions.

The first configuration shows a regular pattern, the second one is generated with the aid of uniformly distributed random numbers and the third one shows clusters. Admittedly, the third configuration is often found in connection with poorly distributed nucleation agents. For the present purpose, however, the second configuration is preferred. It is connected with the simplest point process, the Poisson process, which is based on the lowest degree of information.

As a first step another picture is shown from the original publication. It is introduced here as Fig. 1.22. This picture explains the underlying train of thoughts.

The figure shows five tapered points. From each of these points two curves extend, which are mirrored with respect to a virtual horizontal line drawn through their point of origin x', t'. Let us first consider one pair of curves, which is completely outside the shaded area. The time t' corresponding to the point of departure is the time, when the special spherulite starts growing at x'. Its diameter, which increases with time, is indicated by the increasing stretch in the x-direction between the two curves. In a next step we take point x, t (without accents!). It is maintained

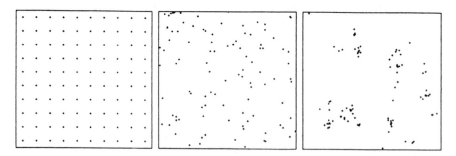

Fig. 1.21 Examples of point configurations on a square: regular, Poissonian, cluster type. [53]. Courtesy of Marcel Dekker, Inc.

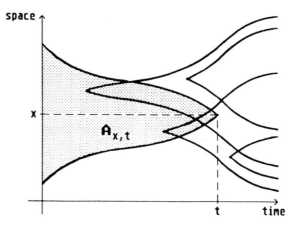

Fig. 1.22 Schematic space-time diagram [52]. The diameters of growing spherulites are given by the stretches on the *x*-axis against time *t*. For an explanation see the text. Courtesy of Marcel Dekker, Inc.

that no spherulite can start at this point. This fact can be explained as follows: Two of the spherulites, which started at earlier times, had their starting points within the shaded area. But this means that the growing spherulites cover the point *x*, *t*. The reader may observe that at every moment all curves have a slope of equal value accept for the sign. These slopes are equal to $\pm G$, if G is the growth speed of all radii at that time (temperature). In emphasizing that this derivation holds for the general case that temperature changes during the growth process, the curves are not drawn as straight lines. In fact, straight lines would be characteristic for the isothermal case with constant G. (G is assumed to be a unique function of temperature, as has been explained earlier in this text.) The shaded area $A_{x,t}$ seems to present kind of an inverse growth area. Apparently, a purely geometric description is given. This means that the density change, which is accompanied by the crystallization, is ignored.

But there is still another point: In reality spherulites extend in three dimensions, in contrast to the one dimension shown in the figure. One can still imagine a presentation in two dimensions. For the purpose, the flat areas between the curves of Fig. 1.22 must be replaced by rotational bodies around the horizontal lines going through the points of departure. Admittedly, three dimensional spherulites cannot be envisaged easily. One has to assume a spherical symmetry in space around the point *x*, *t*, where the time axis is replaced by a series of snapshots. Anyway, one can give a mathematical description for this realistic case.

One looks for a set $A_{x,t}$ of points (x', t') in a four-dimensional space containing time as the fourth variable, where $x(=x, y, z)$ is a vector. For all points of this space the distance between x' and x must be

$$|x - x'| < r(t, t') \tag{1.76}$$

for remaining inside the contour of the mentioned inverse growth area. A point outside this contour cannot give rise to a spherulite covering point *x*, *t*.

In other words, the following set must be found:

$$A_{x,t} = \{(x',t')|t' < t \wedge |x - x'| < \int_{t'}^{t} G(u)\, du \,\}$$ (1.76a)

In a next question it is asked, how large the probability will be that all points of this set are unoccupied by nuclei, which can grow out into spherulites. If no dependence on the location exists, the expectation value for the occupation of each point (volume element) is:

$$d\Lambda = \alpha(t)\, dx\, dy\, dz\, dt$$ (1.77)

where $\alpha(t)$ is the homogeneous rate of nucleation per unit volume of the sample. At this point the distribution function of Poisson has to be consulted. This distribution function has the peculiarity that the probability of the occurrence of an unoccupied point can be given, if the volume element dV is occupied in the average by $d\Lambda$. As is well know, the Poisson distribution reads:

$$\Pi(k) = \frac{\langle n \rangle^k}{k!} \exp\left(-\langle n \rangle\right),$$ (1.78)

where $\Pi(k)$ is the probability for k-fold occupation and $\langle n \rangle$ is the average degree of occupation. If one asks for the probability that $k = 0$, only the exponential is left. In fact, one has $\langle n \rangle^0 = 1$ and $0! = 1$. In our case one obtains:

$$\pi(x,t) = \exp\left[-\Lambda\left(A_{x,t}\right)\right],$$ (1.79)

where Λ is the integral over all $d\Lambda$ of Eq. (1.77). In fact, a sum in the exponent means a product of all individual probabilities of events, which must occur simultaneously. If no restriction by a wall is within reach, i.e. if $r(t, t')$ is not obstructed in its growth by an extra condition, this integral reads:

$$\Lambda(A_{x,t}) = \frac{4\pi}{3} \int_{-\infty}^{t} ds\, \alpha(s) \left[\int_{s}^{t} du\, G(u)\right]^3$$ (1.80)

Actually, the subscript x can be omitted because of the fact that the integral does not depend on any location of the volume elements.

If the lower bound of the first integral is chosen as $-\infty$, this means that the whole crystallization process is included, which starts at a temperature well above the melting point, from where the temperature is continuously lowered to a temperature, where the crystallization proper takes place. If Eq. (1.79) gives the probability that all points of $A_{x,t}$ are unoccupied, the degree of space covering is given by:

$$\xi_g(x,t) = 1 - \pi(x,t) = 1 - \exp\left[-\varLambda(A_{x,t})\right] \qquad (1.81)$$

If $\varLambda(A_{x,t})$ is taken from Eq. (1.80) and the subscript x is omitted, one obtains the well-known equation of Kolmogoroff:

$$\xi_g(t) = 1 - \exp\left(-\frac{4\pi}{3}\int_{-\infty}^{t} ds\ \alpha(s)\left[\int_s^t du\ G(u)\right]^3\right) \qquad (1.82)$$

For a later discussion it is mentioned that

$$\frac{4\pi}{3}\left[\int_s^t du\ G(u)\right]^3$$

is the volume $v(t, s)$ of a sphere, which grows from any location during a time span between s and t.

1.4.2   Behavior of Confined Samples

The question is now, what happens, if one does not look back on hopefully unoccupied sites (volume elements) in a see of liquid but also on unoccupied sites on the surfaces of walls, which confine the sample. For the purpose an infinitely extended slab M of constant thickness D was considered by Eder: $M = [-D/2, +D/2] \times R \times R$. Except for the intensity measure in the fluid, as given by Eq. (1.72), there is now also an intensity measure at the wall surface $\partial M = \{-D/2, +D/2\} \times R \times R$
   One has:

$$d\varLambda_3 = \alpha_3(t)\ dx\ dy\ dz\ dt$$

and

$$d\varLambda_2 = \alpha_2(t)\ dy\ dz\ dt, \qquad (1.83)$$

where the subscripts indicate the dimensions (volume and surface). Also for these equations the condition holds that there is no dependence on the location of the volume or surface elements. With $\alpha_3(t)$ one has the nucleation rate per unit volume of the fluid according to Eqs. (1.34) and (1.35). This means that according to our points of view it is assumed that new nuclei only show up, if the temperature decreases as a function of time. A similar assumption is also made with respect to $\alpha_2(t)$, which is the rate of nucleation per unit surface of the wall.

Instead of Eq. (1.81) one has:

$$\xi_g(x,t) = 1 - \exp\left[-\left(\Lambda_2(A_{x,t}) + \Lambda_3(A_{x,t})\right)\right] = 1 - \exp[-\varphi_0(x,t) - \psi_0(x,t)]$$

$$(1.84)$$

Here the x-coordinate cannot be disregarded. In fact, the x-axis is perpendicular to the wall surfaces. One has:

$$\varphi_0(x,t) = \int_{-\infty}^{t} ds\, \alpha_3(s)\, v(t,s,x)$$

and

$$\psi_0(x,t) = \int_{-\infty}^{t} ds\, \alpha_2(s)\, a(t,s,x) \qquad (1.85)$$

The function $v(t,s,x)$ is the volume of a body, which a sphere with its center at x and a radius $r(t,x)$ cuts out of the sample M. This body may be a full sphere or a sphere truncated on one or on both sides by its intersection with one or both walls. The function $a(t,s,x)$ is the area, which a sphere with its center at x and a radius $r(t, x)$ cuts out of the sample surface ∂M. This area can be zero. But it can also consist of one or two disks cut from one surface or from both surfaces. Table 1.4 gives the pertinent classification.

In this table $R(t) = r(t,-\infty)$ is the radius of a sphere, which in principle starts growing at $t' = -\infty$. This means that, again, the assumption is made that the count of the time starts at a temperature well above the melting point. The temperature is then lowered to a value, where the crystallization proper can occur. But this also means that $G(t)$ is not a constant. It increases continuously from zero to some finite value.

The content of Table 1.4 is illustrated by Fig. 1.23.

In this figure the locations of the classes are shown. The distance between the horizontal lines at $x = \pm D/2$ and the curved lines is $R(t)$. This distance does not increase linearly with time for the reason just explained: Continuous cooling is assumed to be started above the melting point. In class 1 one finds complete spheres and zero areas on the walls. For the use in the other classes two auxiliary functions $\tau_+(t,x)$ and $\tau_-(t,x)$ are required. These functions are defined by the following integrals:

Table 1.4 Classification of the variables (t,x)

Class	Condition
1	$|x| < D/2 - R(t)$
2+	$|D/2 - R(t)| < x < D/2$
2-	$-D/2 < x < -|D/2 - R(t)|$
3	$|x| < R(t) - D/2$

Fig. 1.23 Classification of the variables t and x according to Ref. [22]. Courtesy of Verlag Chemie–John Wiley

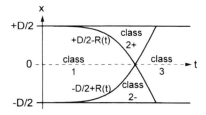

$$\int_{\tau_+(t,x)}^{t} du\, G(u) = \frac{D}{2} - x$$

and

$$\int_{\tau_-(t,x)}^{t} du\, G(u) = \frac{D}{2} + x \tag{1.86}$$

The first function gives the time $s = \tau_+$, at which a spherulite, which is nucleated at $+x$, reaches the upper wall at time t. The second function gives the time $s = \tau_-$, at which a spherulite, which is nucleated at $-x$, reaches the same upper wall at time t. If s is an earlier time than $\tau_+(t,x)$, $v(t,s,x)$ is the volume of a sphere truncated on the upper side and $a(t,s,x)$ is the area of a disk, which is cut from the upper surface. This means that (t,x) is in class 2+. If s is later than $\tau_+(t,x)$, (t,x) is in class 1. However, if s is earlier than $\tau_-(t,x)$, (t,x) is in class 3. As the symmetric case with respect to $x = 0$ is of major interest, class 2- will not be mentioned any more, even if also the asymmetric case with two different surface nucleation rates is within reach.

For the evaluation of these principles straightforward but cumbersome calculations are required. In the review of 1997 [22] the results of these calculations, which have been carried out by G. Eder, are reproduced. Even, if $G(t)$ and $\alpha(t)$, the latter according to Eq. (1.34), allow for a description in terms of a changing temperature, the practical effect is zero. In fact, the function $T(t)$ is not known a priori. It is part of the solution of a heat transfer problem.

For the purpose, a system of rate equations according to the ideas of Schneider et al. [26] would be required. Eder managed to derive those equations for the case of confined samples [22]. He arrived at seven rate equations and two extra differential equations with respect to time for τ_+ and τ_-. Four of the rate equations hold for nucleation in the melt and three for nucleation on the wall surfaces. The structures of these equations strongly depend on the classification given in Table 1.4. So, it does not surprise, if the general case of heat transfer has not yet been put into practice. The merit of Wu´s treatment [35] of heat transfer in DSC will be discussed below. In this treatment Schneider´s original rate equations were used without taking into account the influence of the surface of the pan. First we need a simplified treatment, as given in the next section.

1.4.2.1 Isothermal Case for a Slab of Constant Thickness

For a constant temperature after a sharp quench it is assumed that:

$$\alpha_3(t) = N_V\,\delta_0(t) \quad \text{and} \quad \alpha_2(t) = N_A\,\delta_0(t)$$

together with

$$G = \text{const.} \tag{1.87}$$

In the first pair of equations δ_0 is the Dirac-function concentrated around $t = 0$. At time $t = 0$ a number of N_v nuclei per unit volume comes up suddenly in the bulk and N_A nuclei per unit surface show up suddenly at the surfaces of both walls. Also, the growth speed is constant from the first moment. With respect to the occurrence in the bulk one can say that the validity of the first Eq. (1.87) is not only well-come as a simplification. There are also serious physical reasons for this validity, which will be explained in Chap. 2 of this monograph. A sharp quench, however, is a theoretical assumption.

The results of the pertinent calculations were given in dimensionless coordinates

$$t' = 2\frac{Gt}{D} \quad \text{and} \quad x' = 2\frac{x}{D} \tag{1.88}$$

Two dimensionless parameters were introduced:

$$d_V = \frac{1}{D\sqrt[3]{N_V}} \quad \text{and} \quad d_A = \frac{1}{D\sqrt{N_A}} \tag{1.89}$$

In Fig. 1.24 the results of Eder's calculations are collected.

Various combinations of the parameters d_V and d_A were used. On the vertical axes of the square shaped diagrams the degree of crystallinity is plotted. On the horizontal axes the reduced distance from the mid-plane is drawn. On the right sides one finds the surface of the slab. The reduced times are indicated at the contour lines. Small values of both parameters indicate high number densities of nuclei in the bulk and on the walls. Most interesting are the diagrams at the lower left and at the upper right corners. At the lower left one has a combination of a high number density of nuclei in the bulk and of a low number density of nuclei at the surfaces. Except for the vicinity of the wall one finds a homogeneous increase of the degree of crystallinity with time. Only close to the wall the crystallinity falls off at short times to one half of the value far from the wall. In fact, if the centers of spherulites are at the wall, only half spheres start growing. Initially, these half spheres cover

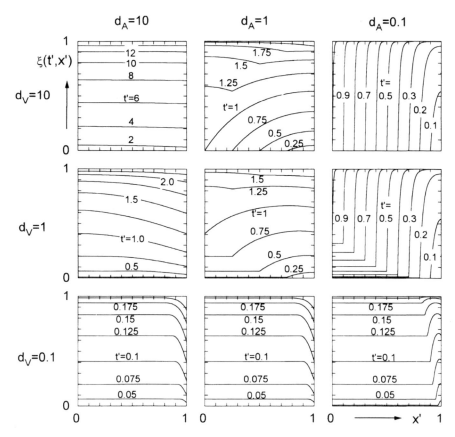

Fig. 1.24 Crystallinity profiles under isothermal conditions for a slab of constant thickness [22]. Results are shown for various combinations of the parameters d and d_A. Courtesy of VCH-Wiley

only half the available space. At the upper right one has an opposite situation: a low number density in the bulk and a high number density at the surfaces. In this case a more or less perfect growth front moves from the surface into the sample. With increasing time the distance from the wall increases, where the crystallinity falls off from one to zero. Nevertheless, one has a dominant effect of transcrystallization. The diagram at the lower right shows the lot of transcrystallization, if there is a high number density of nuclei also in the bulk. Transcrystallization is restricted in this case to a narrow range close to the surface.

1.4.2.2   Significance of the Avrami Index

For an evaluation of the Avrami index one needs the degree of crystallinity averaged over the sample thickness. One has:

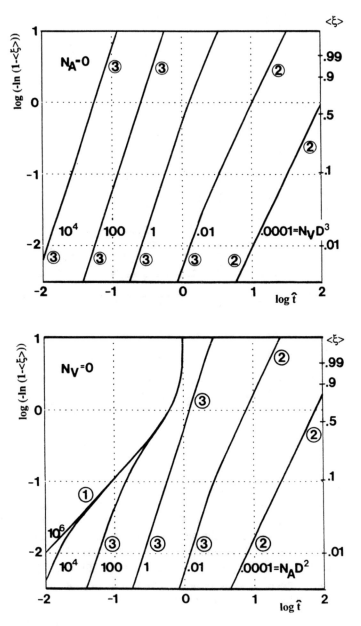

Fig. 1.25 Avrami plots on the basis of the mean degrees of crystallinity for the limiting cases $N_A = 0$ and $N_V = 0$ [52]. The dimensionless parameters $N_V D^3$ and $N_A D^2$, as indicated near the curves from left to right, correspond to the previous parameters of Eq. (1.89) $d_V = 0.0464, 0.216, 1.0, 4.64, 21.6$ and $d_A = 0.01, 0.1, 1.0, 10, 100$, providing a comparison with Fig. 1.24. Courtesy of Marcel Dekker, Inc.

$$\langle \xi_g \rangle (t) = \frac{1}{D} \int\limits_{-D/2}^{+D/2} dx \, \xi_g(x, t) \tag{1.90}$$

For the limiting cases with no nuclei at the surfaces ($N_A = 0$) or in the bulk ($N_V = 0$), corresponding Avrami plots are shown in Fig. 1.25.

The local slopes are given as encircled numbers. The second case is most interesting. One finds a continuous transition from an apparent Avrami index of 2 over an index 3 to an index 1. In fact, at extremely low number densities at the surface one starts with disk-like growth (index 2), passes the number densities, where half spheres are formed (index 3), and ends up with front growth (index 1) at very high number densities on the surface. Both cases, see also the upper figure, show that there is a variety of possible Avrami indices, which can be traced back to the fact that the sample is confined. In an unconfined sample index 3 should be expected with spherulitic growth.

However, confinement does not form the only reason for varying Avrami indices. As mentioned already at an earlier occasion, also the insufficient effectiveness of quenches has its influence. One does not exactly know, when crystallization starts. Eder formulated this problem in a simple way [54]. He defined a dimensionless time with the aid of the time of half conversion $t_{1/2}$. This dimensionless time reads:

$$t^* = \frac{t}{t_{1/2}}. \tag{1.91}$$

It can easily be shown that the Avrami equation reads with this definition of the time:

$$\xi(t^*) = 1 - 2^{-(t^*)^n} \tag{1.92}$$

Now it is assumed that crystallization does not start exactly at $t^* = 0$ but a little later or earlier, because of the uncertainty caused by the badly defined quench. This means that t^* has to be replaced in the above equation by $t^* - t_0^*$ with t_0^* being larger or smaller than zero. In Fig. 1.26 a plot is given of $\log[-\ln(1 - \xi(t^* - t_0^*))]$ versus $\log(t^*)$ for values of t_0^* between -0.3 and $+0.3$, if theoretically an Avrami index of three is expected. On the right ordinate axis also the range of degrees of crystallinity is shown, for which crystallinities can possibly be determined. In this range all curves can easily be approximated by straight lines. With increasing values of t_0^* the slopes of these lines clearly deviate from three. Apparently, ignoring the influence of confinement, one still can obtain deviations from the expected value of the Avrami index.

However, in Chap. 2 of this monograph an extraordinary situation will be described, where the Avrami plot can be applied with success.

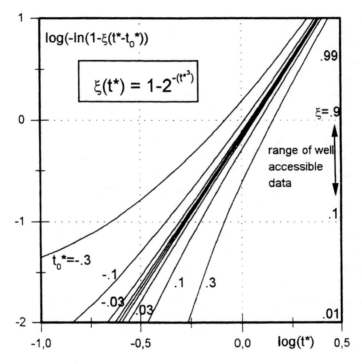

Fig. 1.26 Avrami plots for crystallization processes with different shifted time scales for an expected Avrami index of three [24]. Courtesy of American Soc. Mech. Engng

In this case crystallization is initiated by the application of pressure at a rather high crystallization temperature, where normal crystallization kinetics are still extremely sluggish. The moment, when the pressure is applied, is well defined. And also, one can use rather large samples, so that the confinement is practically of no influence. These conditions form the basis for a successful dilatometry in a special case.

1.4.2.3 Some Final Remarks

After having recognized the importance of the confinement on the kinetics of crystallization one wonders, whether calculations, as carried out at the time by Wu (see Sect. 1.3.3.1), are still of importance. In fact, Wu used the original rate equations by Schneider et al. (see Sect. 1.2.2), which only hold for unconfined media. With these equation she calculated the progress of crystallization under the influence of heat transfer in the pan of a DSC apparatus. Without any doubt, such a

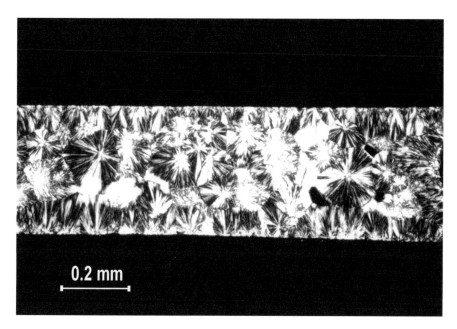

Fig. 1.27 Cross-section from a sample of iPP, as obtained from the DSC pan after a cooling run of 5 K/min [22]. Courtesy of VCH-Wiley

pan forms a serious confinement. In assuaging this situation cross-sections through two samples are shown in Figs. 1.27 and 1.28. These samples come from DSC runs at 5 and 50 K/min. These pictures have been published before [22]. But they are extremely instructive. They show that with iPP the zone, where transcrystallization occurs near the surface, becomes smaller with increasing cooling rate. If a picture is taken, which is obtained after extremely slow crystallization at-say-130 °C, one finds only the footprints of transcrystallization. A look on Fig. 1.1 shows that the number density of nuclei, as found in the bulk of iPP, increases tremendously with decreasing crystallization temperature. (The technique applied avoids surface effects, as will be explained in Chap. 2!) If the number density on the surfaces does not increase as much with decreasing temperature, the thickness of the transcrys-tallized layer must decrease with increasing cooling speed, as the figures actually show. One moves down the diagrams on the right side of Fig. 1.24. Wu´s results gain importance with increasing cooling speeds. If the greater part of the sample crystallized, as if there were no surface influence, one could take the results of Wu´s calculation as significant for the thermal effects envisaged.

Another point of discussion is the unavoidable influence of flow. Actually, there is almost no way to form a material without applying flow. But the influence of flow is particularly effective near the sample surface. And this means that trans-crystallization, as occurring at the surface of a quiescent sample, looses its importance. In fact, as already shown in Fig. 1.1, the number density of nuclei

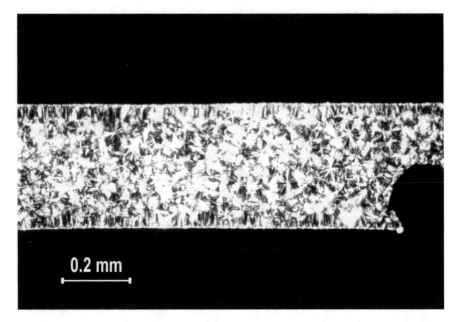

Fig. 1.28 Cross-section from a sample of iPP, as obtained from the DSC pan after a cooling run of 50 K/min [22]. Courtesy of VCH-Wiley

increases so much in the sheared melt (more than fast cooling), that the nuclei on the wall loose their importance anyway (cf. Fig. 1.28).

1.5   Influence of Strong Temperature Gradients

So far, it has always been assumed silently in this report that a spherulite, which starts growing at a certain nucleus, grows in all directions with the same speed, which is determined by the local temperature at the spot, where the nucleus is found. However, if in the vicinity of the nucleus the temperature gradient is strong, this concept fails. The spherulite must grow faster in the direction of decreasing temperature and more slowly in the inverse direction of increasing temperature. If a very fine grained structure is obtained, this effect may be unimportant. In such a case the density of nuclei is high, so that the spherulites impinge before they can grow out into objects of noticeable size. However, not only the temperature dependence of the growth speed must be taken into account. An opposite effect is caused by the fact that the number density of nuclei is not constant but increases with decreasing temperature. In the direction of decreasing temperature one finds an increased speed of growth. At the same time, however, there will be less space for the growth because of the increased density of nuclei. The question is only, which of these two effects will dominate for the final structure in the solidified sample.

If crystallization is stopped by a final quench, before impingement of the growing spherulites occurs, one can expect that, as a first approximation, spherulites remain spherical with unchanged diameter but with an envelope slightly shifted to the side of lower temperatures. It will be difficult to observe such an effect. However, with a sufficiently strong temperature gradient the envelope will become an ellipsoid and its volume will increase. The present section is devoted at least partly to such an effect.

1.5.1   Growth of Spherulites in Temperature Gradient

G.E.W. Schulze and T.R. Naujeck [54] observed a two-dimensional spherulite in a stationary linear temperature field. In such a field the growth speed can be described by $G(x) = G(0)(1 + px)$, where $G(0)$ is the growth speed at the place of the nucleus ($x = 0$). The envelope of a spherulite, which starts growing at $t = 0$ and $x = 0$, has the following shape:

$$(x - x_m(t))^2 + y^2 = r(t)^2 \tag{1.93}$$

with

$$x_m(t) = R(t)g(pR(t)), \quad r(t) = R(t)f(pR(t)) \text{ and } \quad R(t) = G(0)t \tag{1.94}$$

The functions $g(u)$ and $f(u)$ read:

$$g(u) = \frac{\cosh(u) - 1}{u} = \frac{u}{2} + \frac{u^3}{24} + \dots$$

and

$$f(u) = \frac{\sinh(u)}{u} = 1 + \frac{u^2}{6} + \frac{u^4}{120} + \dots \tag{1.95}$$

As already mentioned, the radius of the spherulite does not change to a first approximation because of the behavior of $f(u)$. If $pR(t)$ is very small compared with 1, concentric circles will grow with speed $G(0)$.

Eder generalized these equations for the case that the temperature field is time dependent. He proposed:

$$G(t, x) = G^*(T(t, x)) = G^*(T(t, 0)) \, (1 + p(t)x) \tag{1.96}$$

with

$$p(t) = \frac{\partial \ln G^*}{\partial T}(T(t,0)) \frac{\partial T}{\partial x}(t,0) \tag{1.97}$$

If for some reason the local temperature gradients $\partial T/\partial x$ do not change with the overall temperature (see Sect. 1.3.1, constant cooling rate in slab), $p(t)$ becomes practically constant for any location and for the temperature range, where also $\partial \ln G/\partial T$ is constant (tangent to the right side of Fig. 1.2). Only in the third Eq. (1.94) $R(t) = G(0)t$ has to be replaced by the integral:

$$R(t) = \int_0^t du\, G^*(T(u,0)) \tag{1.98}$$

For the calculation of the degree of crystallinity as a function of time at the position, where the special $\partial T/\partial x$ is envisaged (our x is counted from the place, where the nucleus is located), one can use Eq. (1.81). For the purpose one has only to calculate the proper set $A_{x,t}$. This set consists of all nucleation points (x', t') leading to (deformed) spherulites covering the point of observation (x, t). Instead of the first Eq. (1.76a) one has:

$$(x' - x_m(t',t,x))^2 + (y' - y)^2 \le r(t',t,x)^2, \tag{1.99}$$

where

$$x_m(t',t,x) = x + (1+px)R(t',t)g(pR(t',t))$$

and

$$r(t',t,x) = (1+px)R(t',t)f(pR(t',t)) \tag{1.100}$$

with

$$R(t',t) = \int_{t'}^t du\, G^*(T(u,0)) \tag{1.101}$$

For the validity of Eq. (1.100) it is necessary that x is not too large. One should remain in the vicinity of the origin of the nucleus, so that p does not change too much.

So far, everything is o.k., but for a further calculation some additional assumptions seem necessary. The next simplifying assumption was that the rate of nucleation $\alpha(t', x')$ should be independent of x'. For a calculation of the degree of crystallinity this assumption is reasonable. In fact, the crystallinity depends in first instance on the third power of the growth speed. Only the first power of the number of nuclei is involved. With this simplification the degree of space covering is:

$$\xi_g(t,x) = 1 - \exp\left[-\int_{-\infty}^{t} du\ \alpha(u)\ v(u,t,x)\right]$$

with

$$v(t',t,x) = \frac{4\pi}{3}\ r(t',t,x)^3 \tag{1.102}$$

This radius r can only be bigger than $R(t)$ because of the behavior of function f (u) according to Eq. (1.95). A next assumption was that the rate of nucleation should be concentrated around $t = 0$. ($\alpha(t) = N_V \delta_0(t)$.) With this assumption one has:

$$\xi_g(t,x) = 1 - \exp\left[-\frac{4\pi}{3} N_v\ r(0,t,x)^3\right] \tag{1.103}$$

Finally this equation is capable of the necessary series developments enabling a determination of the tolerable maximum temperature gradient, at which still no observable deviation is found from the derivations of previous sections. With the use of Eq. (1.97) one has (see Eder [22]):

$$\left|\frac{\partial T}{\partial x}\right| \le 3.09 \frac{\sqrt[3]{N_v}}{\left|\frac{\partial \ln G}{\partial T}\right|} \sqrt{\varepsilon} \tag{1.104}$$

In this equation ε is the permitted relative error for the temperature gradient. If $\varepsilon = 0.05$, one has for iPP with $(1/G)\partial G/\partial T = 0, 2\ K^{-1}$ values for the tolerable temperature gradient, which still depend on the number density of nuclei. For $N_V = 10^{12}\ m^{-3}$ and for $N_V = 10^{15}\ m^{-3}$ (see Fig. 1.1) one obtains 3×10^4 K/m and 3×10^5 K/m, respectively. In practical processing, however, those gradients are often exceeded.

In this connection one should not forget, however, that with respect to the morphology of the end-product the ideas that the rate of nucleation or the number density of nuclei are independent of the x-coordinate, seem incorrect. In fact, with high temperature gradients the number density of nuclei increases tremendously with decreasing temperature, so that the space for growth in that direction is reduced. This effect counterbalances the just described effect. Unfortunately, a pertinent theory does not yet exist. It will also be difficult to carry out a decisive experimental study. But the existence of such a counter-balancing effect can be demonstrated at the other end of the "spectrum".

Fig. 1.29 Morphology at the crystallization zone of an iPP, if this zone has proceeded over a considerable distance into the sample. The fine grained structure on the right is caused by the final quench. For conclusions see the text. Ewa Ratajski, unpublished

1.5.2   Counter-Balance by a Gradient in the Number Density of Nuclei

In connection with this problem the reader is reminded of Fig. 1.13. In this figure the progress of the diffuse crystallization zone is followed as a function of time for a high density polyethylene up to a distance of about 3 mm from the quenched wall. In the apparatus, which is used for the purpose, a cylindrical polymer sample of a diameter of 2 cm has been quenched on one end surface to the temperatures indicated in the graph. The temperature of the other end surface was kept at the original temperature of the sample of 170 °C. The distance between the end surfaces is 10 mm in this apparatus. A similar experiment was also carried out with a sample of iPP. (See Fig. 1.14.) In principle, the temperature of the diffuse zone increases with its distance from the quenched wall. But from Fig. 1.2 we know that above 140 °C the number density of nuclei is immeasurably small in iPP. (An explanation of this fact will be given in Chap. 2.)

In the present Fig. 1.29 a special morphology of a sample of iPP is shown at a distance of 2.6 mm from the other end surface, which was kept at 180 °C (after a preheating of the whole machine to 240 °C). The quenched surface stayed at 120 °C. It took days, before the diffuse crystallization zone reached this distance of 7.4 mm from the quenched wall. The sample as a whole was quenched after that long time to a lower temperature for a fixation of the obtained morphology. As one can see on the figure, an almost sharp growth front was found at the moment of the quench. This growth front was possible because of a lack of nuclei on the side of the higher temperatures. There was enough space for an undisturbed though slowed down growth. One can see also that at lower crystallization temperatures, which prevailed a little earlier, i.e. on the lee side of the growth front, a lot more nuclei influenced the picture. This pronounced increase of the number density with decreasing temperature is in accord with Fig. 1.1. Looking at this figure we also have asked the legitimate question, how low the number density must be at temperatures above-say-140 °C-in a quiescent melt of iPP. These figures fit together in a convincing way.

Remains now a discussion of the vicinity of the growth front. One does not need much fantasy to imagine the completed shapes of the big outgrowths. One probably would have found big ellipsoids with their focuses near the colder side. One of these focuses can even be sensed. But there is no doubt that the growth speed is higher at the colder side. However, growth is apparently hampered much more at that side by the larger number of nuclei. On the warmer side the growth speed is certainly lower, but this drawback is outweighed by the reduced number of nuclei. These conclusions remain valid even if the faster growing β-crystal-modification is involved, as may be suggested by the bigger birefringence of the outgrows [44]. In this connection also the experiences of Lovinger et al. [45] with enclosures of the other modification should not be overlooked.

References

1. Tribout C, Monasse B, Haudin JM (1996) Experimental study of shear-induced crystallization of an impact polypropylene copolymer. Colloid Polym Sci 274:197–208
2. Janeschitz-Kriegl H, Ratajski E, Stadlbauer M (2003) Flow as an effective promotor of nucleation in polymer melts: a quantitative evaluation. Rheol Acta 42:355–364
3. Stadlbauer M, Janeschitz-Kriegl H, Eder G, Ratajski E (2004) New extensional rheometer for creep flow at high tensile stress, part II. Flow induced nucleation for the crystallization of iPP. J Rheol 48:631–639
4. Marand H, Xu J, Srinivas S (1998) Determination of the equilibrium melting temperature of polymer crystals: linear and non-linear Hoffman-Weeks extrapolation. Macromolecules 31:8219–8229
5. Janeschitz-Kriegl H (1983) Polymer melt rheology and flow birefringence. Springer, Berlin, pp 181–187, p 424
6. Smoluchowsky M (1916, in German) An attempt of a mathematical theory of the kinetics of coagulation in colloidal solusions. Z phys Chemie XCII:129–168
7. Ratajski E, Janeschitz-Kriegl H (1996) How to determine high growth speeds in polymer crystallization. Colloid Polym Sci 274:938–951
8. Gandica A, Magill JH (1972) A universal relationship for the crystallization kinetics of polymeric materials. Polymer 13:595–596
9. Janeschitz-Kriegl H, Eder G, Stadlbauer M, Ratajski E (2005, in English) A thermodynamic frame for the kinetics of polymer crystallization under process conditions. Monatshefte für Chemie 136:1119–1137

10. Van Krevelen DW (1990) Properties of polymers, 3rd edn. Elsevier, Amsterdam, Oxford, New York, Tokyo, p 120
11. Woodward AE (1989) Atlas of polymer morphology. Hanser Publishers, Munich, Vienna, New York, p 106–109
12. Prime RB, Wunderlich B, Melillo L (1969) Extended chain crystals V. Thermal analysis and electron microscopy of the melting process in polyethylene. J Polym Sci A-2 7:2091–2097
13. Keller A (1957) Single crystals in polymers: evidence of folded-chain configuration. Philos Mag 2:1171–1175
14. Hoffman JD, Miller RL (1997) Kinetics of crystallization from the melt and chain folding in polyethylene fractions revisited: theory and experiment. Polymer 38:3151–3212
15. Eder G, Janeschitz-Kriegl H, Krobath G (1989) Shear induced crystallization, a relaxation phenomenon in polymer melts. Progr Colloid Polym Sci 80:1–7
16. Azzurri F, Alfonso GC (2005) Lifetime of shear-induced crystal nucleation precursors. Macromolecules 38:1723–1728
17. Blundell DJ, Keller A, Kovacs AJ (1966) A new self-nucleation phenomenon and its application to the growing of polymer crystals from solution. Polym Lett 4:481–486
18. Kolmogoroff AN (1937, in Russian) On the statistical theory of the crystallization of metals. Bull Acad Nauk SSSR Math Ser 1:355–359
19. Avrami M (1939–1941) Kinetics of phase change I, II, III. J Chem Phys 6:1103–1112, 8:212–224, 9:177–184
20. Evans VB (1945) The laws of expanding circles and spheres in relation to the lateral growth rate of surface films and grain size of metals. Trans Faraday Soc 41:365–374
21. Tobin MC (1974, 1976) Theory of phase transition with growth site impingement I, II. J Polym Sci Phys Ed 12:399–406, 14:2253–2257
22. Eder G, Janeschitz-Kriegl H (1997) Processing of polymers: crystallization. Mat Sci Tech (VCH-Wiley) 18:269–342
23. Wunderlich B (1973) Macromolecular physics, vol 1, p 282. Academic Press, New York, London
24. Eder G (1997) The role of heat transfer problems in standard crystallization experiments. ASME Int HTD 351:131–137
25. Nakamura K, Watanabe T, Katayama K, Amano T (1972) Some aspects of non-isothermal crystallization of polymers. I relationship between crystallization temperature, crystallinity and cooling conditions. J Appl Polym Sci 16:1077–1091
26. Schneider W, Köppl A, Berger J (1988) Non-isothermal crystallization of polymers. Intern Polym Proc 2:151–154
27. Van Krevelen DW (1978) Crystallinity of polymers and the means to influence the crystallization process. Chimia 32:279–294
28. Eder G (1998) Crystallization in polymer processing: modelling and experimentation. In: Alkeryd L et al (eds) Progress of industrial mathematics at ECMI 98. Teubner, Stuttgart, Leipzig 1999, p 138
29. Janeschitz-Kriegl H, Eder G (1984) A less familiar feature of crystalline layer growth on a cold surface. Plast Rubber Process Appl 4:145–148
30. Astarita G, Kenny JM (1987) The Stefan and Deborah numbers in polymer crystallization. Chem Eng Commun 53:69–110
31. Janeschitz-Kriegl H, Eder G, Ratajski E (2006) A process classification number for the solidification of crystallizing materials. Intern Polym Proc 21:521–526
32. Van Antwerpen F, Van Krevelen DW (1972) Influence of crystallization temperature, molecular weight and additives on the crystallization kinetics of poly(ethylene terephthalate). J Polym Sci Polym Phys Ed 10:2423–2435
33. Pijpers TFJ, Mathot VBF, Goderis B, Scherrenberg RI, Van der Vegte EW (2002) High-speed calorimetry for the study of the kinetics of (de)vitrification, crystallization and melting of macromolecules. Macromolecules 35:3601–3613

34. Janeschitz-Kriegl H, Wippel H, Paulik Ch, Eder G (1993) Polymer crystallization dynamics, as reflected by differential scanning calorimetry. Part I: on the calibration of the apparatus. Colloid Polym Sci 271:1107–1115
35. Wu CH, Eder G, Janeschitz-Kriegl H (1993) Polymer crystallization dynamics, as reflected by differential scanning calorimetry. Part II: numerical simulations. Colloid Polym Sci 271:1116–1126
36. Eder G, Janeschitz-Kriegl H, Liedauer S (1990) Crystallization processes in quiescent and moving polymer melts under heat transfer conditions. Progr Polym Sci 15:629–714
37. Adamovsky SA, Minakov AA, Schick C (2003) Scanning microcalorimetry at high cooling rates. Thermochim Acta 403:55–63
38. Adamovsky S, Schick C (2004) Ultra-fast isothermal calorimeter using thin film sensors. Thermochim Acta 415:1–7
39. Minakov AA, Mordvintsev DA, Schick C (2004) Melting and reorganization of poly(ethylene terephthalate) on fast heating (1000 K/s). Polymer 45:3755–3763
40. Minakov A, Morikawa J, Hashimoto T, Huth H, Schick C (2006) Temperature distribution in thin-film chip utilized for advanced nanocalorimetry. Meas Sci Technol 17:199–207
41. De Santis F, Adamovsky S, Titomanlio G, Schick C (2006) Scanning nanocalorimetry at high cooling rate of isotactic polypropylene. Macromolecules 39:2562–2567
42. Janeschitz-Kriegl H (1996) The role of transport phenomena in polymer science. J. Macromol Sci-Pure Appl Chem A 33:841–858
43. Berger J, Schneider W (1986) A zone model of rate controlled solidification. Plast Rubber Process Appl 6:127–133
44. Turner-Jones A, Aizlewood JM, Beckett DR (1964) Crystalline forms of isotactic polypropylenes. Makromol Chem 74:134–158
45. Lovinger AJ, Chua JO, Gryte CC (1977) Studies of the α and β forms of isotactic polypropylene by crystallization in a temperature gradient. J Polym Sci Polym Phys Ed 15:641–656
46. Carslaw HS, Jaeger JC (1959) Conduction of heat in solids, 2nd edn. Clarendon Press, Oxford, p 285
47. Eder G, Janeschitz-Kriegl H (1984) Stefan problem and polymer processing. Polym Bull 11:93–98
48. Janeschitz-Kriegl M, Janeschitz-Kriegl H, Eder G, Forstner R (2006) Heat transfer through metal walls of finite thiskness. Intern Polym Proc 21:41–48
49. Bundrup J, Immergut H (1990) Polymer Handbook. Wiley
50. Ullmann F (1990) Enzyklopaedie der techn. Elsevier, Chemie
51. Dittus FW, Boelter LMK (1930) Heat transfer in automobile radiators of tubular type. Publications on Engineering, Berkley
52. Eder G (1997) Fundamentals of structure formation in crystallizing polymers. In: Natada K, Kitayama T, Vogl O (eds) Macromolecular design of polymeric materials. Marcel Dekker Inc., pp 761–782
53. Daley DJ, Vere-Jones D (1988) An introduction to the theory of point processes. Springer, New York
54. Schulze GEW, Naujeck TR (1991) A growing 2D spherulite and calculus of variation. Colloid Polym Sci 269:689–695

Chapter 2
Kinetics and Structure Formation in Unloaded Quiescent Melts

2.1 Introductory Remarks

It goes without saying that structure formation in a permanently quiescent unloaded melt does hardly ever occur in practical polymer processing. In fact, flow and pressurization are virtually inevitable. Nevertheless, the present chapter will prove to be of great importance, as flow or pressure induced crystallization cannot be understood without a profound basic knowledge of the processes occurring in a permanently quiescent melt, which has not been put under pressure. Such a melt must be cooled down in its quiescent state from a temperature well above the equilibrium melting point, where the residues of previous crystallization processes are erased.

2.2 Empirical Techniques

2.2.1 Number Density of Nuclei

The number density of nuclei is of particular interest for the structure formation. This quantity determines the number density of spherulites, which is finally obtained and which determines various properties of the product. In fact, a fine grained structure is preferable in many cases because of the good ductility obtained with such a structure. On the other hand, as has been pointed out in Sect. 1.2.1, the growth speed of the spherulites mainly determines the speed of the solidification process and, as a consequence, the heat transfer problems.

From the early times up to now there has been a discussion, whether the assumption of sporadic nucleation with a temperature dependent rate of nucleation or just the assumption of a predetermined number density of nuclei as a unique function of temperature will be most appropriate for a description of the

© Springer International Publishing AG 2018
H. Janeschitz-Kriegl, *Crystallization Modalities in Polymer Melt Processing*,
https://doi.org/10.1007/978-3-319-77317-9_2

solidification process in quiescent polymer melts (see the alternative use of Eqs. (1.10) or (1.11)). The use of the latter assumption and equation has been advocated by Van Krevelen in an early paper [1] just because of the easier handling. However, fortunately it has turned out meanwhile that there are strong physical reasons for the latter assumption. One can be quite sure that below the melting temperature of the spherulites (see Fig. 1.3) all nuclei are of the athermal type [2, 3], which means that they have their specific temperature of activation. As a consequence they become effective immediately, when this temperature is reached during the cooling process. A theoretical discussion will be given below. In the present section just the experimental methods will be described, which lead to the determination of the said number densities as functions of temperature.

Janeschitz-Kriegl et al. [3] described a successful method for the determination of the number density of an industrial PP as a function of temperature. For the purpose a cylindrical sample of a diameter of 4 mm was prepared in the solid state. In the heart of this sample a thin hole was drilled from one end down to half the length of the sample. In this hole a thin thermocouple with a diameter of 0.3 mm was placed, so that its junction was at the end of the hole. This sample was wrapped into a metal foil and suspended in a horizontal position in the coil of a wire. The whole was placed into a glass cylinder, which could be closed at both ends. A heating wire was wound around its outer surface. This glass tube had two inlets near its ends and an outlet in the middle. In the beginning there was no fluid in the tube. The sample was heated by radiation to a temperature well above the equilibrium melting point. In this way remnants of previous crystallization were erased. After the moment, when the radiation heater was switched off, a fast stream of a fluid of a temperature much lower than the intended crystallization temperature, was let in through one of the inlets. In this way the sample was cooled much faster than in the case that from the beginning a fluid of the intended temperature would have been admitted. However, at the moment, when the thermocouple in the heart of the sample passed the intended temperature, a fluid of this temperature was introduced through the second inlet. It replaced the colder fluid in a counter-current. In this way the cooling was quite abruptly stopped. The number density of nuclei is assumed to be equal to the number density of spherulites, which is found after the final solidification of the sample. It goes without saying that this solidification has to occur at the temperature, for which the determination of the number of nuclei is intended. A good measure for this density is the number of spherulites sectioned per unit surface of the cross-section, after being raised to the power 3/2. For this procedure part of the cross-section is used, which is obtained close to the heart of the sample, sufficiently remote from the sample surface. The difference in bulk density of the melt and the solidified sample is ignored in this method. In view of the large effects of temperature changes on the number density of nuclei, however, this ignorance has no serious consequences.

Cooling curves, as obtained with this type of experiments, are shown in Fig. 2.1. Average cooling rates are given near the curves. Always a certain undershoot is found for the temperature. For the evaluation the lowest temperature was used. One of the reasons for this choice is that nuclei, which show up at a certain temperature, will quickly be stabilized so that a mild re-heating will not change their number. But such a re-heating, if it should be of influence, could only reduce the number of nuclei. Also, if cooling is not fast enough, some nuclei will start growing already during the cooling cycle. The rising spherulites will cover some space, where otherwise further nuclei can come up. But this means that, in principle, the said method can give only an underestimate of the number density. The scope of such an underestimate can only increase with decreasing crystallization temperature.

This statement is of particular interest in view of the enormous increase of the measured number density of nuclei, as found with decreasing temperature. This means that, in reality, this increase, which is shown in Fig. 2.2, can only be more pronounced. For the microscope it requires a remarkable selection of magnification optics.

The curve of Fig. 2.2 has been transferred to the left side of Fig. 1.1. As already mentioned, when this latter figure was introduced, a logarithmic scale had to be used for the number density because of its enormous increase with decreasing temperature. For i-PP a range of temperatures from about 110 down to 85 °C was covered. Previously, a range of similar scope has never been envisaged.

A similar result was obtained according to Braun et al. [4] with a sample of PB-1, which did not contain a nucleation agent (PB0300). Another sample, which was nucleated (PB0110), showed a larger number density over the whole range of temperatures from about 110 to 50 °C. This difference was particularly pronounced

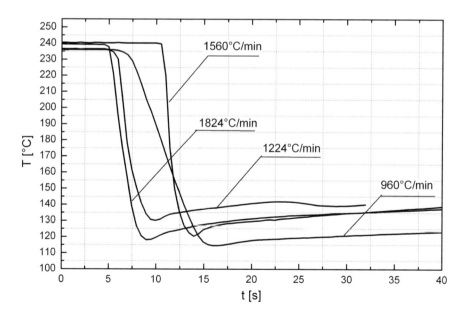

Fig. 2.1 Sample temperatures as functions of time for samples of an industrial PP, according to the described method. Average cooling rates are indicated near the curves [3]. Courtesy of Springer Verlag

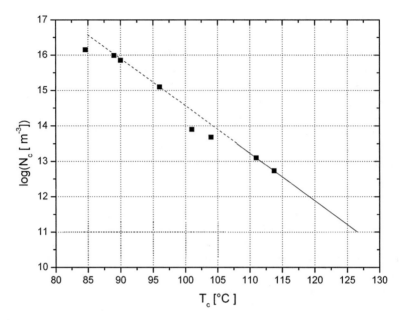

Fig. 2.2 Number density of nuclei as a function of crystallization temperature for a sample of industrial PP according to [3]. Full line obtained on samples from the DSC-machine (see Figs. 1.27 and 1.28) with the use of Eq. (1.67) for the temperature. The dotted line is the extrapolation to lower temperatures. Black squares for KS 10/new method. Courtesy of Springer Verlag

at the higher temperatures. This is in accord with the intended influence of a nucleation agent. In Fig. 2.3 the pertinent results are shown.

Two polyketones of different degrees of copolymerization were investigated by Stadlbauer et al. [5]. Accept for the just described counter-current method also a thin layer technique was applied. In the latter method a very thin sample of a thickness of a few micrometers is prepared by evaporating the droplet of a dilute solution on a cover glass. The obtained layer is then covered by a second cover glass and reheated under slight pressure to a temperature well above the equilibrium melting point for obvious reasons. The application of a thin layer has the advantage that the number of spherulites, which is finally found, is tremendously reduced and can easily be counted. This fact has been advocated by Chew, Griffiths and Stachurski [6]. In fact. only those spherulites can grow in such a layer, which have their nuclei within the volume of this layer. And the surface of the glasses has no influence on the number of nuclei, because glass is an undercooled fluid. If one takes a thin slice, as cut from the bulk of an already solidified sample, this slice contains also parts of spherulites, which have their nuclei outside the volume of this slice.

In Fig. 2.4 the said thin slice apparatus is shown [5]. The two chambers of this apparatus are separated by the sandwich of the sample. Two heating blocks, which

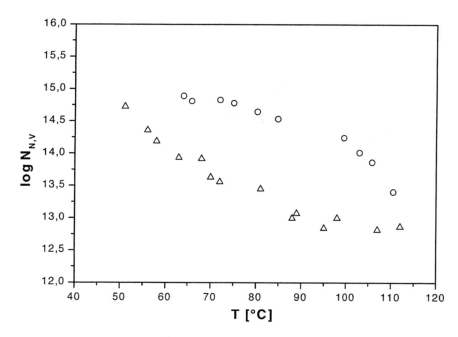

Fig. 2.3 Number densities (per m³) against crystallization temperatures for PB0110 (circles) and PB0300 (triangles) [4]. PB0110 was nucleated. Courtesy of the Society of Plastics Engineers

are located above and below the chambers, serve for the initial heating erasing remnants of previous crystallization. When the experiment proper is started, two streams of a heat transfer fluid, which are at the intended crystallization temperature T_c, enter the chambers. Because of the fact that the sample is extremely thin, there is no possibility to measure the sample temperature itself. But a thermocouple is located as close as possible to the sample. One has to trust that the sample temperature adjusts quickly enough to the fluid temperature as the thermocouple does. The course of the temperature of the thermocouple is shown in the lower part of the figure. After some time, which may be called the crystallization time t_c, streams of lower temperature can replace the original streams. These streams cause an almost abrupt decline of the temperature. This trick is of particular importance, if the growth speed of the spherulites is to be determined. This will be explained below.

For the present purpose an early interruption of the crystallization will keep the formed spherulites separated from each other. In fact, with the polyketones the samples can be quenched into the glassy state at 15 °C, so that the surroundings of the spherulites can be kept clear. (With i-PP one obtains a fine grain phase, which can easily be discerned from the spherulites.) A separation of the spherulites is important, if fast growing spherulites must be counted in a system of high nucleation density. In fact, those spherulites are still small, when impinging, and their boundaries are not well developed. This will be shown to be the case with polyethylenes.

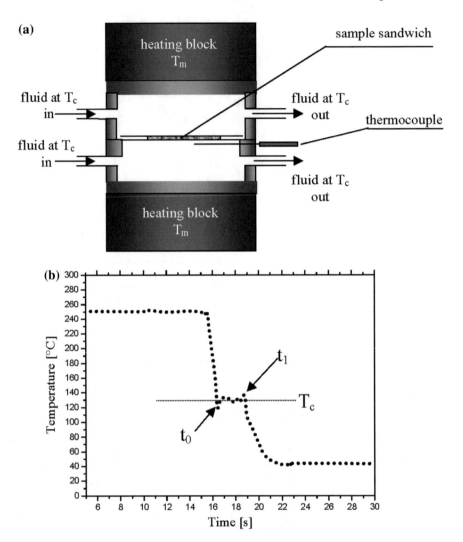

Fig. 2.4 The thin slice machine according to Stadlbauer et al.   [5]. In the lower part the temperature profile is shown for $T_c = 130$ °C and $t_c = 2$ s. Courtesy of Elsevier

The number of spherulites, which are counted per unit surface under the microscope, must not be raised to the power 3/2 in this case. One must just divide this number by the thickness of the sample, in order to obtain the number density of nuclei per unit volume. The accuracy of this method depends in first instance on the accuracy of the thickness measurement. Nevertheless, for the polyketones a comparison with the results of the counter-current method turns out to be quite satisfying. This is shown in Fig. 2.5.

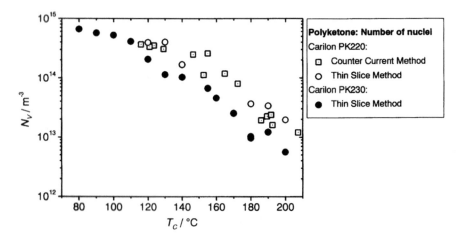

Fig. 2.5 Number density of nuclei versus crystallization temperature for two polyketones of Basell with 6% propylene (PK230) and with 3% propylene (PK220) as replacements for ethylene in the copolymer of carbon monoxide and ethylene. Full circles for PK230 as obtained in thin slice measurements. Open circles for PK220 from thin slice measurements. Open squares for PK220 from counter-current measurements [5]. Courtesy of Elsevier

The number densities of PK220 are a factor four larger than those for PK230. Interestingly, the melting points of the spherulites show an inverse trend: For PK220 one has $T_m = 220$ °C and for KP230 one has $T_m = 230$ °C. The polymer, which does not contain a fraction of propylene, has a melting point of 257 °C, which is too high for convenient processing.

Finally, the problems with polyethylene will be discussed. The thin slice method has just been described. In Fig. 2.6 a photograph is shown of a sample of HDPE of a thickness of 0.85 µm, as obtained by M. Stadlbauer. This sample was quenched from 180 to 100 °C. The growth of the spherulite was stopped by a second quench with ice water, before coalescence of the (disk-like) "spherulites" occurred. A number of 300 spherulites was counted on an area of 7.13×10^{-8} m^2. Knowing the thickness of the sample one arrives at 4.5×10^{16} nuclei per cubic meter.

The result of the count is given in Fig. 2.7, where also the number densities of nuclei of other industrial polymers, as investigated in our group, are shown for comparison. For the investigated HDPE the number densities of nuclei increase with decreasing temperature down to a temperature of about 80 °C. With lower temperatures smaller values were found, which cannot easily be explained. In fact, with all the other polymers the number densities of nuclei continuously increase with decreasing temperature, in accordance with Eq. (1.41), as must be expected for athermal nuclei (see the consideration given below). Remarkably, the steepness of the increase is particularly high with i-PP. For HDPE and PET the numbers are particularly high over the whole range of temperatures.

As just mentioned, the kink in the curve for HDPE cannot easily be explained. But in this respect one should not forget the difficulties experienced with these

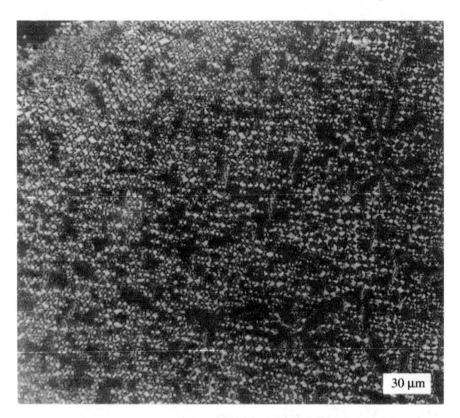

Fig. 2.6 Micrograph of a layer of HDPE of a thickness of 0.85 μm after a quench from 180 to 100 °C and a second quench with ice water just before the spherulites could coalesce. Obtained by M. Stadlbauer . Courtesy Springer Wien

measurements, when the temperature is further lowered. Apparently, the 100 °C of Fig. 2.6 were still high enough for avoiding those difficulties. Anyway, E. Ratajski investigated a series of copolymers of polyethylene with increasing amounts of propylene. With these polymers the number densities of nuclei decreased. But below 70 °C she was not able to count the spherulites. It is not only that the number of rising spherulites increases. Also their contours become less sharp.

The results presented in this section will serve as a basis for the understanding of the nature of athermal nuclei in polymers and for the influence, which flow can exert on the number density of effective nuclei and on the formation of thread-like precursors ("shishs") leading to highly oriented structures. These facts explain, why the determination of the number densities of nuclei is treated here so extensively. So far there are no investigations of other laboratories, which have been extended over such large ranges of temperature. In this connection the reliability of our measurements has been questioned. However, the fact that two quite different types of measurements (raising the number of the sectioned spherulites to the power 3/2 and

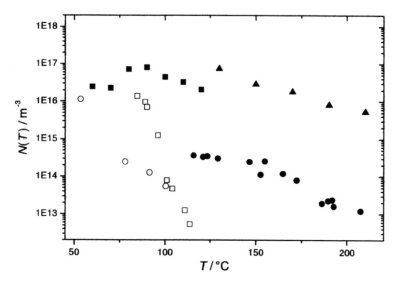

Fig. 2.7 Number densities of nuclei for several industrial polymers as functions of temperature: full squares HDPE, open squares i-PP, open circles PB-1, closed circles polyketones, closed triangles PET, according to [7]. Courtesy Springer Wien

counting spherulites in a thin slice) led to comparable results (see Fig. 2.5), should convince the reader.

2.2.2 Growth Speeds of Spherulites

This subject has tradition in the famous work by Magill and his coworkers, of which the paper by Gandica and Magill is quoted here [9] as a representative contribution. These authors investigated the growth speeds of spherulites of a great number of polymers, which were prepared in the laboratory. All these polymers showed growth speeds, which were not too high, so that the growing spherulites could be observed under the microscope. The growth speeds of all polymers showed a maximum halfway between the melting point and the glass-transition temperature. This fact is in contrast to the temperature dependence of the number densities. As we have seen, these densities increase continuously with decreasing temperatures. The authors arrived at the conclusion that a universal curve could be constructed, if the ratio G/G_{max} is plotted against a dimensionless temperature Θ, where G_{max} is the growth speed at the maximum and the definition of Θ reads as follows:

$$\Theta = \frac{T - T_\infty}{T_m - T_\infty},$$

(2.1)

where $T_\infty = (T_g - 50 \text{ K})$, with T_m being the melting point and T_g the glass transition temperature. In this connection the reader should be reminded of the remarks made, when Eq. (1.40) was introduced for an approximate description of the temperature dependence of the growth speed with respect to the roles of T_m and T_g as limiting parameters for the growth range.

2.2.2.1 Growth Speeds of Fast Growing Spherulites

In general the growth speeds of industrial polymers are too high for an observation under the microscope, if the interesting temperature ranges are envisaged. The microscope can be used only at temperatures rather close to the melting point. But at those temperatures solidification occurs at a slow pace, which is of no interest for practical processing. In order to cover the whole range of the dimensionless temperatures of Fig. 2.8, samples must be rigorously quenched. In fact, as will be shown below, with polymers of increasing growth speed also the distance between the melting point and the glass transition temperature increases. A special paper has been devoted to the methods of determining high growth speeds by Ratajski and Janeschitz-Kriegl [10].

Fig. 2.8 Dimensionless master curve for the growth rates ($v = G$) of a variety of synthetic polymers according to Gandica and Magill [9]. Courtesy of Elsevier

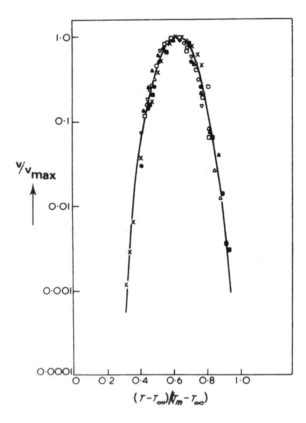

   The most direct method is the quench of ultra thin slices. For the purpose slices of 5 μm are cut with a microtome from a solid sample and embedded between cover glasses (of 0.15 mm thickness). These samples were heated in a first stage to a temperature well above the equilibrium melting point in order to destroy residues of previous crystallization. In a next step they were quenched with a heat transfer fluid to proper temperatures of crystallization. At those crystallization temperatures the samples are kept for various time spans (crystallization times), before being quenched to a much lower temperature, where crystallization is arrested. With some polymers this can be a temperature below the glass transition temperature, for others only a temperature, where fast crystallization leads to a fine grained matter. This matter separates the spherulites formed at the higher crystallization temperature. For the first quench diethylene glycol has been used in most cases. The second quench occurred with ice water, in which the ethylene glycol dissolves. The solidified samples were put under the microscope, where the radii of the spherulites, which actually are disc-like because of the low thickness of the samples, are measured. If these radii are plotted for equal crystallization temperatures against the crystallization times, one always obtains straight lines. The slope of such a line gives the speed of growth at the chosen temperature. For i-PP the results of this investigation are shown in Fig. 2.9.

   Some remarks have to be made with respect to Fig. 2.9. First of all it is remarkable that the radii of the obtained spherulites do not differ very much from each other. This fact is explicitly shown in Fig. 2.10. This uniformity points to the

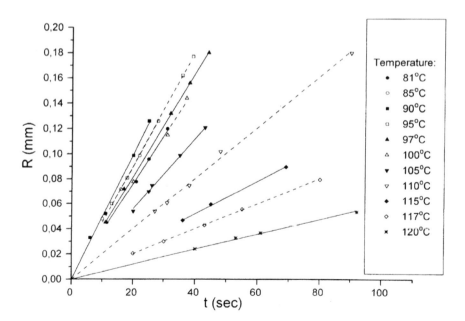

Fig. 2.9 Radii of the biggest spherulites of an industrial PP against the temper times for a series of bath temperatures [10]. Courtesy of Springer Verlag

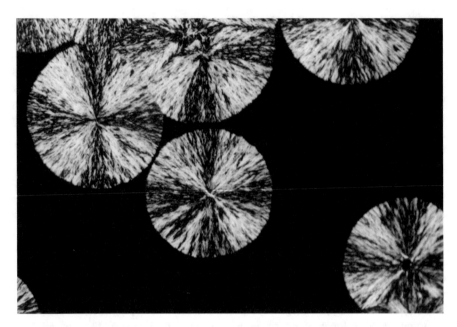

Fig. 2.10 Micrograph of two-dimensional "spherulites" of an industrial PP, as obtained in a bath of 95 °C after a temper time of 39 s [10]. The diameter of the biggest spherulite (upper left corner) is 0.177 mm. Courtesy of Springer Verlag

fact that the bulk of the spherulites starts growing practically at the same time. Probably only few of them started a little earlier before the temperature, at which the quench aimed, had exactly been reached. This fact underlines our assumption of the occurrence of athermal nuclei, which have their exact temperature of sudden activation. But the uniformity is also of practical use.

In fact, after each quench one finds other spherulites in contrast to the microscope, where the growth of the same spherulite can be followed. From Fig. 2.9 one also learns that there is practically no incubation time. Within the accuracy of the measurements all lines start at the origin. By the way, with PB-1 a retarded start was found at a temperature closer to the melting point (at 100 °C), when growth of spherulites was investigated under the microscope [4]. At the lower temperature of 71 °C, where the above technique was applied, the line started at the origin, as in Fig. 2.9 [5]. There are also examples, where the straight line started at a positive intercept with the ordinate axis. This happens with fast growing spherulites, if the quench is not fast enough but sufficiently reproducible. This happened for HDPE at temperatures below 85 °C [7].

Another point is that the observed spherulites are actually disc-like. If the contact angle between the spherulite, the melt and the glass is finite, which must be assumed, the growth speed can only be reduced with respect to the growth speed of a free spherulite in the melt. However, the effect cannot be very large, as the consistency of the spherulite, which contains still a lot of amorphous material, cannot differ very much from that of the pure melt. Also, PP shows polymorphism [11]. Interestingly enough, however, only spherulites of the α-modification were observed in this investigation. These spherulites show a low birefringence in contrast to the spherulites of the β-modification. This contrast will be subject of discussion in one of the next paragraphs. Apparently, the β-modification is nucleated only at metal walls or by the addition of a proper nucleation agent. (Fortunately, glass walls do not nucleate any modification. In fact, glass is an undercooled fluid.)

In our laboratory also other methods were applied in order to determine growth speeds. It goes without saying that the application of various techniques is important for a check of the reliability of determinations. Short descriptions will be given of two techniques: Transcrystallization against a temperature gradient and onset of turbidity after touch with a thermostated wall.

As already pointed out at earlier occasions in this monograph, transcrystallization [12] is hampered regularly by the occurrence of spherulites at some distance from the wall. This happens particularly, if the initial temperature of the melt is not too high and remains uniform at some distance from the quenched wall. This effect can be avoided, if the sample thickness is chosen small enough, so that the second confining wall, which is kept at the original high temperature, keeps its influence on the temperature distribution in the sample. In such a case transcrystallization occurs against a steep temperature gradient. Admittedly, such a gradient causes a retardation of the front movement, but reduces the probability for the occurrence of spherulites. The pertinent experiment occurs in two steps. In the first step one wall is quenched to the desired crystallization temperature. After a chosen time both walls are quenched to a much lower temperature, where a fine grained structure is obtained during continued crystallization. Finally, the solidified sample is removed from the confinement and a cross-section is made in a plain perpendicular to the plain of the confining walls. In this cross-section the location of the undisturbed crystallization front can easily be observed.

For i-PP, however, which has been our appropriate experimental object, such experiments cannot be carried out directly because of the polymorphism of this polymer. The consequences of this polymorphism are shown in Fig. 2.11. This figure shows large parts of an undisturbed growth front of the α-modification, as observed under the polarizing microscope. Unfortunately, there are places at the wall, where the β-modification is nucleated. As this modification grows faster, one obtains a bright cone with its apex at the point at the wall, where this modification is nucleated. (The cross-section does not go exactly through this point.) The brightness of this cone is due to the high birefringence of the β-crystallites. Those phenomena were observed first by Lovinger et al. [13]. On Fig. 2.11 it can be seen easily that at a certain distance from the wall the growth front becomes one of the β-modification. (In the beginning of our investigations we overlooked this fact. It led to an overestimate of the growth speed of the α-modification in a temperature range between 90 and 110 °C [14], see Fig. 6 of this paper, or [15], Fig. 3.)

In avoiding this uncertainty a rigorous choice was made. The wall was rubbed in with Cinquasia Gold (Permanent Red E3B), a nucleation agent for the β-modification.

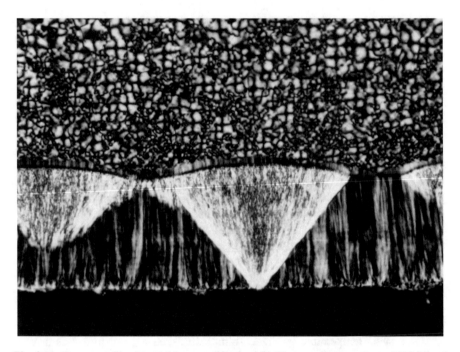

Fig. 2.11 Transcrystalline layer of the α-modification of i-PP at a wall temperature of 110 °C and a contact time of 103 s. A single β-nucleus at the wall produces a conical body, which becomes dominant after a distance of 0.153 mm from the wall [10]. Courtesy of Springer Verlag

In this way an undisturbed crystallization front of the β-modification was obtained. Such a front is already shown in Fig. 1.15 of the first chapter. In the present Fig. 2.12 the propagation of this type of fronts is shown for a series of quenched wall temperatures at a temperature of the opposite wall of 200 °C. The initial slopes of these curves should give the growth speeds of β-spherulites at the indicated temperatures of the quenched wall. However, the determination of those initial slopes is extremely inaccurate. At this point a method, as proposed by G. Eder [8], has been introduced.

Eder's method analyzes the shape of the curves over a bigger distance from the wall. To start with, it was shown that the critical Fourier number was passed over in all experiments. This means that the temperature profile should be linear, if the production of latent heat was ignored. The Fourier number reads:

$$Fo = \frac{at}{D^2}, \qquad (2.2)$$

where a is the heat diffusivity of the melt, t is the time and D is the distance between the walls. For $a = 10^{-7}$ m^{2}s^{-1} and $D = 1$ mm one obtains from Eq. (2.2) a critical time $t = 5$ s. A look at Fig. 2.12 shows that all points were found at larger times. Two assumptions were made for the required calculations (see also [8]):

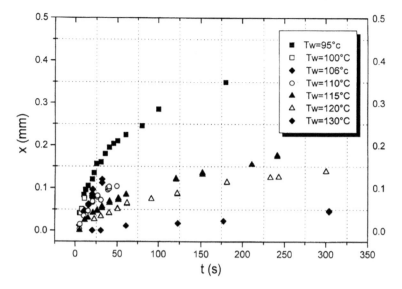

Fig. 2.12 Front positions of the transcrystalline β-layers as functions of time for various temperatures of the quenched wall. The temperature of the opposite wall at a distance of 1 mm was 200 °C [10]. Courtesy of Springer Verlag

(a) For a sufficiently steep temperature gradient the kink, as caused by the latent heat, can be ignored. So one has for the temperature profile:

$$T(x) = T_w + (T_i - T_w)\frac{x}{D}, \tag{2.3}$$

where x is the distance from the quenched wall and T_i is the initial temperature of both walls (i.e. the temperature of the opposite wall) and T_w is the temperature of the quenched wall after the quench.

(b) For the growth speed as a function of temperature the following equation has been used:

$$G(T) = G_w \exp\left(-\beta(T - T_w)\right) \text{ with } G_w = G(T_w) \tag{2.4}$$

The reader will notice a contrast between this equation and Eq. (1.40), where a quadratic dependence on temperature is proposed. But T_w is not T_{max} and Eq. (2.4) describes a tangent to log(G) versus T at some temperature T_w along this curve. With these assumptions the differential equation for the progress of the growth front reads:

$$\frac{dx}{dt} = G(T(x,t)) = G_w \exp\left(-\beta \frac{T_i - T_w}{D} x(t)\right).$$ (2.5)

This equation can easily be integrated. One obtains:

$$\log(x(t)) = \log(G_w) + \log\left[\tau \ln\left(1 + \frac{t}{\tau}\right)\right].$$ (2.6)

$$\text{with} \quad \tau = \frac{D}{\beta(T_i - T_w)G_w}.$$ (2.6a)

It may be recognized that τ depends on the location along the curve log (G) versus T, i.e. on T_w—also through β, T_w and G_w. If for a series of assumed values of τ a pattern of curves $\log[\tau \ln (1+t/\tau)]$ versus $\log(t)$ is prepared, this pattern covers all combinations of the parameters of Eq. (2.6a) within the mashes of this pattern. Such a pattern has been drawn on a transparent sheet and put on the experimental curve $\log(x(t))$ versus $\log(t)$ and shifted along the log x -axis until the shape of one of the curves of the pattern fitted the shape of the experimental curve. In this way the ordinate of the corresponding $\log(G_w)$ was obtained with reasonable accuracy. In Fig. 2.13 the result of this procedure is given for a large number of wall temperatures between 80 and 120 °C for the growth speed of β-spherulites of i-PP. Only at higher temperatures growth speeds of these spherulites were known from the work of Lovinger et al. [13].

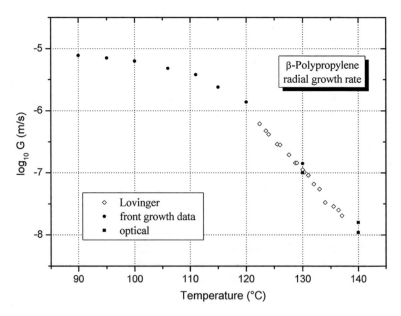

Fig. 2.13 Growth speeds of β-spherulites of i-PP as functions of temperature according to Eder's method for temperatures between 80 and 120 °C [10]. At higher temperatures growth speeds, as obtained by Lovinger et al. [13], are taken over. Courtesy of Springer Verlag

In a next step Lovinger's cone angle K was used in order to obtain the growth speeds of the α-spherulites in the same range of temperatures (see Fig. 2.11). In fact, one has:

$$\frac{G_\alpha}{G_\beta} = \cos\left(\frac{K}{2}\right). \tag{2.7}$$

The points, which were obtained in this way, fitted nicely into the picture reproduced in Fig. 1.2 together with points, which were obtained in the thin slice experiments. Only at higher temperatures many points of other authors could be included. So far in the present section results were reported, which were obtained with transcrystallization against a steep temperature gradient.

Interestingly enough, the second method announced above, i.e. onset of turbidity after touch with a thermostated wall, is based on a quite different physics. It is based on the time needed for a crystallization process starting at time zero until it becomes apparent with the aid of light scattering. In this respect the reader is reminded of interesting work by Berger and Schneider [16]. For the case that no nucleation takes place at the quenched wall, the said authors showed that a zone of diffuse crystallization must move into the melt of originally uniform temperature.

If one considers the line of half conversion, this line has a shape very similar to the line of the square root law. The only grave difference is that it does not start at $x = 0$ but is shifted downwards the x-axis, so that it cuts a positive section from the time axis. Of course, negative distances x are only formal. The zone starts at the said positive value of the time, a fact which creates the impression of an incubation time. However, no incubation time is in the basic assumptions of the model. A look on Fig. 1.14 of Chap. 1 conveys the experimental situation. In fact, also the experimental curves start at positive values of time. So, there must be another reason than incubation for the delay in response. And there is only one way to explain this phenomenon: light scattering becomes intensive enough only after some time. Initially spherulites or their precursors have a scattering power proportional to the square of their volume. They are Rayleigh scatterers. As the volume of a ball shaped scatterer is proportional to $(Gt)^3$, one obtains for the development of the scattering power S of the unit of volume of the sample:

$$S = C N (Gt)^6, \tag{2.8}$$

where C is an optical constant, N is the number of nuclei per unit volume, G is the growth speed of the spherulites and t is the time elapsed from the moment of the quench of the wall. Equation (2.8) presents a parabola of the sixth order, which shows an upswing only after a considerable time. During this time span the crystallization process remains unobservable. In reality a more complex scattering mechanism can be developed. However, such a scattering will not occur earlier than the one predicted by Eq. (2.8). If now also the assumption is made that the development of structure follows the same route, independent of the speed of formation, one obtains the following equation for the growth speed:

$$G = \left(\frac{S}{C}\right)^{1/6} \frac{1}{N^{1/6}t}. \tag{2.9}$$

Anyway, one can conclude from this equation that the time elapsing until the moment, when the light scattering becomes observable, must be inversely proportional to the growth speed, whereas the number density of nuclei can only play a minor role. In fact, the sixth root of this quantity cannot change very much within the range of temperatures, where this equation will be of use. Obviously, Eq. (2.9) must be calibrated at a higher temperature, where the parameters G and N can be determined along other routes. Preferably, a comparison over a larger range of temperatures should be accomplished. This was no problem with the quite slowly crystallizing PP.

An excuse for the application of this equation is that it can be used, if the time span becomes so short that other techniques can no longer be applied. For many years this was the case for HDPE at temperatures remote from the melting point. At time zero sample melts were brought into contact with a metal wall, which was thermostated at the required crystallization temperature. Shortly after this moment the contact surface was abruptly clad by a film of opaque material. This process could even be followed with the aid of a stop watch. In a more accurate method the moment was determined, when the reflection of a light beam was interrupted. This light beam was directed through the still transparent melt on the polished contact surface. Fractions of a second could be determined in this way. Only more recently, M. Stadlbauer [7] improved the thin slice technique, so that a comparison with his more direct measurements was possible.

As an example Fig. 2.14 is introduced. It shows the growth speed of spherulites of HDPE as a function of temperature. Down to 123 °C the results of Chew et al. [6] are reproduced. These results were obtained by direct measurement of spherulite radii (black squares). Our Eq. (2.9) was calibrated at 120 °C with the result of these authors. In this way the range of measurements could be extended by more than 40 °C to lower temperatures. Down-pointed triangles stand for the results obtained with interrupted reflection. Upright triangles are obtained with the stop watch. Open circles give results obtained from the intercept of the diffuse crystallization zone with the time axis. (With HDPE the zone of transition from the clear melt to the homogeneously opaque material is very narrow in contrast to the zone obtained with i-PP, which is much broader, cf. Figs. 1.13 and 1.14.)

The theoretical line, which is shown in Fig. 2.14, will be explained in Sect. 2.3.1.

Stadlbauer was able to extend the range of measurements with the improved thin slice method on both sides of the range of Chew et al., namely to lower and also to higher temperatures. These results are given in Fig. 2.15, which shows a survey of all growth speed measurements, which were carried out in our laboratory. With HDPE one notices that at about 80 °C there is a downwards kink in the curve. We suspect that measurements below this temperature are no longer reliable. Above this temperature Stadlbauer's points exactly fall on the line of Fig. 2.14. This fact

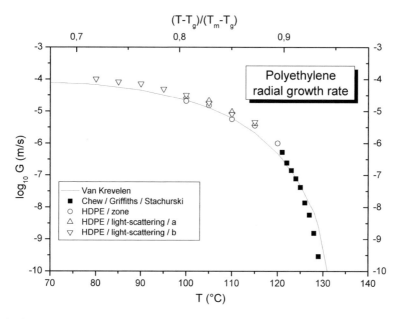

Fig. 2.14 Growth speeds of spherulites of HDPE over a wide range of temperatures. Black squares stand for measurements in a thin slice equipment [6]. Other points were obtained with the use of Eq. (2.9) [10]. Courtesy of Springer Verlag

endorses the usefulness of Eq. (2.9). In the graphs, where radii of spherulites are plotted against temper time for temperatures below the kink, the straight lines, as drawn through the points, do not go through the origin. (Cf. Fig. 2.9.) Positive intercepts with the ordinate axis are found. This shows that the quenches were not fast enough for preventing premature spherulite growth. In this connection one should not forget that the cover glasses are rather thick compared with the slices.

On Fig. 2.15 one can see the enormous differences in the growth speeds of diverse industrial polymers. In fact, similar differences are shown in Fig. 2.7 for the number densities of nuclei. In particular, the number density of nuclei shows a particularly strong increase with decreasing temperature for i-PP. For HDPE the level of the number density of nuclei is much higher, but less dependent on temperature. But this is in favor for the application of Eq. (2.9). $N^{1/6}$ will be practically independent of temperature.

Finally it must be emphasized that the value of Eq. (2.9) is more in the field of general insights. The usefulness of this equation shows that there are no incubation times. Also the simultaneity of the start of the growth for all nuclei is of importance for our insight into the physics. At temperatures below the melting temperature of the spherulites one finds only nuclei, which are "prefabricated" by the so-called "local alignments". This will be explained in Sect. 2.3.2.

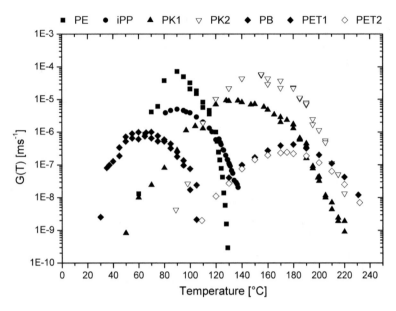

Fig. 2.15 Growth speeds of several industrial polymers as functions of temperature: full squares HDPE, full circles iPP, full diamonds PB-1, full and open triangles two polyketons of different comonomer content, open diamonds PET of two molar masses, according to [7]. Courtesy of Springer Wien

2.2.2.2   Growth Speed and Small Angle Light Scattering

The intention to determine the growth speed of spherulites with the aid of small angle light scattering (SALS) dates back to the sixties. At that time Stein and Rhodes [17] published a trend setting paper, in which they showed that a scattering pattern of the shape of a four-leafed clover is obtained, if crossed polars (H_V) are used. In this pattern the distance of the center of light intensity in the leafs from the point of impact of the primary beam is inverse proportional to the spherulite radius. The pertinent theory was improved step by step, until Keijzers [18] found a quite simple equation for this relation. This equation reads:

$$R = \frac{4.1\,\lambda}{4\pi\,\sin(\Theta_m/2)},\qquad(2.10)$$

where R is the radius of an ideal spherulite, λ is the wave length of the light and Θ_m is the scattering angle, both in the medium. An ideal spherulite is a spherulite, in which the polarizable elements are in an exactly radial position (polarization only radial and tangential). An underground scattering is added for the imperfections. Keijzers published this equation only in his thesis [18] and not in the pertinent paper with Van Aartsen and Prins [19]. But Van Antwerpen and Van Krevelen used this equation for their investigation of the growth speeds of spherulites of PET

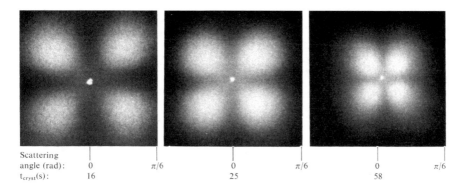

Scattering angle (rad):	0	$\pi/6$	0	$\pi/6$	0	$\pi/6$
t_{cryst}(s):	16		25		58	

Fig. 2.16 H_V scattering patterns for a PET sample of a number average molar mass of 35400 Dalton. Crystallization temperature is 175 °C. Crystallization times 16, 25 and 58 s. The polarization directions are vertical and horizontal. These results are from refs. [20, 21]. Courtesy of the American Chemical Society

[20, 21]. These authors also checked the equation with nylon, which forms beautiful spherulites. Unfortunately, one needs a light-scattering photometer in order to follow the development of the pattern during the growth process. Van Antwerpen monitored the light intensity along a line under 45° with respect to the polarizer.

Some representative results of Van Antwerpen and Van Krevelen are shown in Fig. 2.16. In this figure the four-leafed clover is shown for three stages in the development of the crystallization of a PET sample.

It goes without saying that also some explicit results must be presented. For the purpose Fig. 2.17 is introduced. In this figure results are combined, which were obtained by heating the sample from the glassy state and by cooling the sample from the melt to the required crystallization temperature. No significant differences were found between the results of both methods. A regression was applied for the calculation of the drawn lines. Only for three samples the measurement points were inserted. In contrast to the experience with samples of sufficiently high molar masses a strong dependence on the molar mass was obtained for these low molar mass samples. A discussion of this contrast will be given in the next section.

A continuation of the discussion on the usefulness of SALS will be given in Chap. 3 in connection with flow induced crystallization. A reason for the better chances of this method in flow experiments is the fact that flow induced crystallization can be investigated at much higher temperatures, where practically nothing happens in unloaded quiescent melts. A consequence of this fact is that the samples can be cooled without much troubles to the experimental temperatures, before flow is started. Also pressurization, which enhances crystallization, can be activated after such a relatively slow cooling process. In an unloaded quiescent melt, however, the sample must be cooled to much lower temperatures for the initiation of interesting phenomena. For rather slowly crystallizing materials like i-PS or PET there is still no big problem with cooling. However, for fast crystallizing polymers the situation changes drastically. As an example the behavior of HDPE is quoted. When using

Fig. 2.17 Spherulite growth
rates versus crystallization
temperatures for PET samples
of different molar masses M_n
in Daltons: **1: 19 000, 2: 22
300, 3: 24 000, 4: 27 000, 5:
30 100, 6: 34 500, 7: 39 100,**
according to [21]. courtesy of
John Wiley

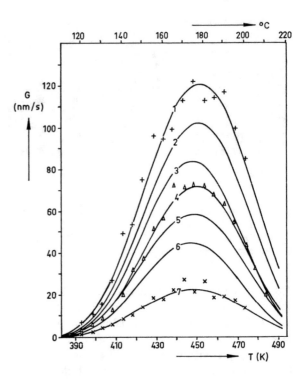

Eq. (2.9), which certainly is loathed by scattering specialists, one encounters times
below one second with quiescent HDPE. I do not think that the above technique can
be applied under those extreme conditions.

2.3   Theoretical Considerations

2.3.1   Theory of the Growth Speed of Spherulites

In this section the work of Van Krevelen [1, 22] will provide the guideline. The
results of this work are of high practical value, and there is no reason to believe that
these results have become obsolete. However, as will be shown in the next section,
there are good reasons, why the classical approach, as applied by Van Krevelen, is
valid only for the growth speed, but not for the primary nucleation. In fact, the
growth of lamellae, of which spherulites are built, just requires local rearrangements
of parts of neighboring macromolecules. (See the independence of the growth speed
of the molar mass, except for very short molecules [23].) In contrast, major rearrangements should be required for the formation of primary nuclei, if the surrounding would be at random. But the topology of long molecules with their
surroundings prohibits those undertakings effectively. We were forced to accept this

hitherto uncommon vision because of the results of our experiments. This fact will make Sect. 2.3.2 particularly thrilling. However, the local mobility, which can be proved even for some materials in the glassy state (see e.g. Heijboer [24]), seems to preserve the classical theory of the growth speed.

After this introduction it will be clear, why only the equation for secondary nucleation, which is responsible for the growth mechanism, is quoted here. One has for the rate of growth:

$$G = G_0 e^{-\frac{E_D}{RT}} e^{-\frac{\Delta G_2}{kT}}. \tag{2.11}$$

In this equation G_0 is a constant of the order of kT/h, E_D is an activation energy for the transport per mol of a designated part of a molecule and ΔG_2 is the free energy of a local nucleus for two-dimensional growth. Interestingly, the last term of this equation gives a Boltzmann distribution for the occurrence of this state of energy. Actually, such a Boltzmann term only holds for an equilibrium situation, which is not given here. The first exponential term serves as a substitute for this omission. Interestingly enough, Ziabicki and Alfonso [25] have shown quite recently that a correct description of the "growth diffusion" against the slope of the barrier ascending to ΔG_2 can be given in terms of a suitable Fokker-Planck equation. An absorbing wall at the top serves as the adequate boundary condition. Unfortunately, the authors were not able to give the resulting flux over the barrier in terms of measurable parameters.

Van Krevelen [1] derived a semi-empirical equation for the activation energy occurring in the first exponential (the transport factor). For the purpose he evaluated elder results by Mandelkern et al [26]. In terms of the characteristic temperatures (T_m and T_g) of a variety of "normal" polymers Van Krevelen's equation reads:

$$\frac{E_D}{R} = C_D \frac{T_m^2}{T_m - T_g} \tag{2.12}$$

with $C_D \approx 5$.

A little earlier Hoffman [27] proposed the use of the WLF-equation [28] for a calculation of the transport factor for lower temperatures (more close to the glass transition), where a temperature independent activation energy no longer holds. This equation reads:

$$\frac{E_D}{RT} = \frac{C_1}{R(C_2 + T - T_g)} \tag{2.13}$$

with the "universal" constants $C_1 = 17.2$ kJ/mol and $C_2 = 51.6$ K.

In this way one arrives at a rather simple semi-empirical expression for the growth rate of spherulites:

$$\ln(G) = \ln(G_0) \; - \; \frac{E_D}{RT} \; - \; \frac{265\text{K}}{T} \frac{T_m}{\Delta T} \tag{2.14}$$

with $G_0 = 10^{12}\,\text{nm/s}$.

One can now introduce Eq. (2.12) or Eq. (2.13) into Eq. (2.14). In carrying out this substitution two dimensionless variables $\xi = T_m/T_x$, with T_x being the crystallization temperature, and $\delta = T_g/T_m$ are introduced. Also a certain temperature T_k is defined, which lies between T_g and T_m and will be scrutinized a little later. One obtains for $T_x \geq T_k$:

$$\log G = \log G_0 - 2.3 \frac{\xi}{1-\delta} - \frac{115\text{K}}{T_m} \frac{\xi^2}{\xi-1}. \tag{2.15}$$

For $T_x < T_k$ one has:

$$\log G = \log G_0 - \frac{895\text{K}\,\xi}{51.6\text{K}\,\xi + T_m(1-\delta\,\xi)} - \frac{115\text{K}}{T_m} \frac{\xi^2}{\xi-1}. \tag{2.16}$$

In Fig. 2.18 both equations have been used. For the right side Eq. (2.15) is responsible. The corresponding curve has been calculated up to the highest point, which defines the said temperature T_k. In a next step Eq. (2.16) has been used for a calculation of the lower part on the left side. As the temperature range for the validity of this equation is restricted, this curve has been connected by hand with the apex of the right curve. In this way a most interesting result is obtained. Dependent on the value of the parameter $\delta = T_g/T_m$ one obtains curves with their maxima at the specific value of T_k. But the heights of these maxima differ enormously, i.e. from almost 10^6 nm/s to about 10^{-16} nm/s, if δ varies between 0.45 and 0.90. Admittedly, these curves only hold for so-called normal polymers. However, the general behavior of polymers follows the trend reflected in this picture: the closer the glass transition temperature is to the melting point, the lower is the growth speed. Also, the picture of Fig. 2.18 holds for a melting point of 473 K. For every 10 °C, that T_m is higher or lower than 473 K, $\log(G)$ (G is replaced by v in the graph) will be about 0.1 higher or lower than given in the graph.

By the way, most satisfactory is the observation that, apparently, HDPE belongs to the "normal" polymers. In fact, in Fig. 2.14 a theoretical line is shown, which agrees rather well with the experimental points. This line is obtained with the aid of Eq. (2.15), if the following values are inserted for the characteristic temperatures: $T_m = 144$ °C ($= 417$ K) and $T_g = -84$°C ($= 189$ K). As was pointed out, when Fig. 2.15 was discussed, the kink in the curve for HDPE should be ascribed to the uncertainties of the thin slice measurements below 80 °C. Above this temperature, there is a very good agreement with the values shown in Fig. 2.14. But this means that the theoretical curve can replace the experimental points below 80 °C. So, it seems clear that the kink is an artifact.

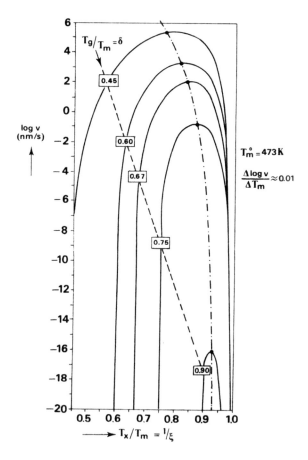

Fig. 2.18 Master curves for the linear growth speeds of spherulites (v instead of G) versus dimensionless crystallization temperature T/T_m (= $1/\xi$), with $T_g/T_m = \delta$ as parameter, according to Van Krevelen [1, 22]. Courtesy of Brunner, Zürich, and of Elsevier

Quite different is the behavior of i-PP. Van Krevelen's curve does not fit the experimental points for the α-modification (see Fig. 1.2). Between 100 and 150 °C it lies above the experimental points with the biggest distance of one decade at about 130 °C. Only the maximum growth speed is correctly predicted. Within the range of usual tacticities these tacticities do not seem to play a predominant role. And the spherulites of the β-modification play their own game: Over the whole range of temperatures the growth speeds of the β-spherulites are higher than those of the α-spherulites. This happens not withstanding the fact that the temperature, where the spherulites of the α-modification are melting, is higher than the corresponding temperature for the β-modification. Details can be found in ref. [10], where also the references are given.

The described deviations of the growth speed of the α-modification of PP from the so-called "normal" behavior have certainly to do with the fact that the α-modification of PP shows several growth regimes, as has been described by Hoffman et al. (See [29, 30].)

Finally, it should be mentioned that Van Antwerpen [31] has investigated more closely the molar mass dependence of the growth speed, as it has been found for low molar mass samples. According to this author this dependence is contained in the second term. One has:

$$G_{0,M} = G_0 + \frac{\text{const.}}{M_n}, \tag{2.17}$$

where M_n is the number average molar mass. Obviously, $G_{0,M}$ goes to the constant parameter G_0, if M_n becomes large enough, as it should be according to the general experience. The present author is permitted to say that he has difficulties with this interpretation. Apparently, the higher growth speeds of those low molar mass samples are due to enhanced transport mechanisms. Probably, this discrepancy has to do with the problem addressed by Ziabicki and Alfonso [25].

For more recent developments in the field one can consult the second edition of Mandelkern's book [32] or a paper by Hoffman and Miller [33]. For our purpose, however, the results of Van Krevelen's school must be preferred, because of their greater practical relevance. In the book by Mandelkern a host of results is given, which surpasses the scope of the present treatise, which aims at structure formation. Unfortunately it must be said that with respect to the development of the theory Mandelkern's presentation is not completely up to date. For instance, Tobin's theory is treated to the full extent. But in Sect. 1.2.1 of the present treatise it is shown that this theory is a theoretical artifact. Its deviation from Kolmogoroff is caused only by an unnecessary fictitious simplification. In fact, the same model is used. A similar criticism holds for the theory by Nakamura et al. Mandelkern apparently overlooked that Nakamura's treatment only holds, if the number density of nuclei is independent of temperature and predetermined. Nakamura's theory cannot hold for the general case. Amongst others this becomes evident from Fig. 1.1 of the present treatise, a figure which unfortunately was only published about the same time as Mandelkern's second edition. The criticism on the said theories was stressed already in our review from 1997 [8], seven years before Mandelkern's second edition appeared. A valid alternative to Nakamura's theory was already published in 1988 by Schneider et al. (see Sect. 1.2.2 of the present treatise). This theory is not mentioned by Mandelkern.

2.3.2 On the Nature of Primary Nuclei in Polymer Melts

2.3.2.1 Some Comments on the Theory by Hoffman and Lauritzen

In two pioneering papers [34, 35] these authors opened the discussion of the nucleation in polymer crystallization. In their theoretical approach they obtained lengthy nuclei. In fact, this type of nuclei must be expected, if long molecules are associated. But the assumption, which was made by the authors in order to obtain those nuclei, was not directly the presence of long molecules but the occurrence of an increased surface tension at the end surfaces with respect to the side surfaces. Actually, it took us quite a time before we realized this fact. Apparently, the authors exchanged cause and consequence. Our biggest riddle was in this respect the self-nucleation effect, as discovered by Blundell, Keller and Kovacs [36].

Many people have tried meanwhile to reproduce the self-nucleation effect, which had been found for the first time by the just mentioned authors. In fact, if a carefully grown spherulite is slowly heated under the microscope up to a temperature, where it just melts, one can find quite a lot of new nuclei within the domain of the old spherulite, when the temperature is lowered again. This is the self-nucleation. If the spherulite is heated a little less carefully, one can get back at least the old spherulite at the same place. But what does this effect mean? In the latter case it meant that the nucleus, on which the spherulite had grown for the first time, survived the melting of the spherulite. In the case of self-nucleation it is even shown that many new nuclei, which came into existence in the previous crystallization process, survived at a temperature, where the spherulite melted. So far, nobody dared to explain this self-nucleation. A big crystalline body melts, but tiny remainders survive. By accident the author of this monograph met a person from the sugar industry, who provided the priming.

2.3.2.2   The Theory by Larson and Garside

Studying the behavior of saturated sugar solutions Larson and Garside [37, 38] discovered a serious shortcoming of the classical theory of nucleation, a theory which goes back to Becker and Döring [39]. In this classical theory the assumption is made that the barrier ΔG in the free energy between small enclaves of the new phase and the continuum of the fluid mother phase is a consequence of the surface tension existing between the phases. And the additional assumption was made that this surface tension is independent of the size of the enclave of the new phase. Along this route the mechanism of sporadic primary nucleation has been derived. One obtains for the free energy difference ΔG_r of a spherical enclave:

$$\Delta G_r = \frac{4\pi}{3} r^3 \frac{\rho}{M} \Delta\mu + 4\pi r^2 \sigma. \tag{2.18}$$

In this equation r is the radius of the new phase, ρ is its density, M is the molar mass, $\Delta\mu$ is the difference in the chemical potential of the new and the old phase and σ is the surface tension.

The first derivative with respect to r is zero at $r = 0$ and at the critical value:

$$r_m = \frac{2\sigma M}{|\Delta\mu|\rho}, \tag{2.19}$$

where ΔG_r has its maximum

$$\Delta G_{r,m} = \frac{16\pi}{3}\frac{\sigma^3 M^2}{(\Delta\mu)^2 \rho^2}. \tag{2.20}$$

Only, if r is larger than r_m, such a nucleus can grow, because only then ΔG_r decreases with increasing r. It is important to realize that particles of the new phase are formed sporadically. And if $r < r_m$, they disappear again as a consequence of the thermal motion. But Eq. (2.20) only works if

$$\Delta G_{r,m} \geq kT. \tag{2.21}$$

In fact, if $\Delta G_{r,m}$ is smaller than kT, such a barrier becomes ineffectual. For $\Delta\mu$ the well-known first order approximation in terms of the undercooling ΔT can be introduced. It reads:

$$\Delta\mu = \frac{\Delta H_{tr}}{T_{tr}}\Delta T, \tag{2.22}$$

where ΔH_{tr} is the latent heat of the phase transition and T_{tr} is the equilibrium temperature of this transition. If Eq. (2.22) is introduced into Eq. (2.20), one obtains:

$$\Delta G_{r,m} = \frac{16\pi}{3}\frac{\sigma^3 M^2 T_{tr}^2}{\rho^2 \Delta H_{tr}^2}\frac{1}{(\Delta T)^2}. \tag{2.23}$$

This equation reminds us of Eq. (2.14) except for the fact that the square of ΔT shows up in the teller, because of the three-dimensional case represented here.

In Fig. 2.19 the shape of the curve, as described by Eq. (2.18), is reproduced as a full line of ΔG versus r. It is characterized by a slope zero at $r = 0$. However, Larson and Garside pointed to the fact that the said surface tension is not a constant, but decreases with decreasing size of the newly formed particle. This fact has been stressed already quite early by Tolman [40]. Tolman showed that the transition zone of thickness δ between the two phases is responsible. As a consequence, in the second term of Eq. (2.18) σ must be multiplied by $f(\delta/r)$, a function, which goes to zero, if the ratio between brackets goes to infinity, i.e. if r goes to zero. Rusli and Larson [41] chose an exponential function. Unfortunately, the parameters of such a function are not known. Nevertheless, qualitatively one obtains the result, which is shown by the dashed line in Fig. 2.19. The conclusion must be that small enough associates should be stable, a fact which could not be explained by the classical theory. Colloid chemists may be grateful for this conclusion. Larson and his collaborators could show that clusters of sugar molecules, but also of ionic substances, are stable in still clear super-saturated solutions.

Fig. 2.19 Gibbs free energy
of a spherical phase against its
radius according to [41].
Courtesy of Pergamon Press

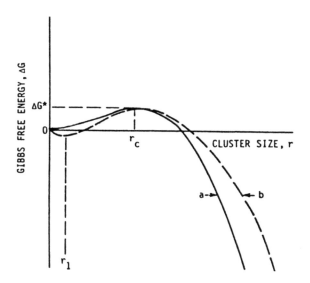

The implications of this result for an explanation of self-nucleation will be
treated in the next section. In prevision of its importance the work of Larson et al.
was already quoted in our early review article from 1990 [15].

By the way: From the upper transition in Fig. 1.3 one can estimate an equation
for the maximum tolerable undercooling ΔT [42]. One obtains:

$$\Delta T = \frac{1}{3} \frac{\Delta H_{tr}}{C_{p,l}}, \tag{2.24}$$

where $C_{p,l}$ is the specific heat of the fluid. One can show that this equation is not
unrealistic. For the purpose one has to put $\Delta G_{r,m}$ on the left side of Eq. (2.23) equal
to kT and fill in the value for ΔT from Eq. (2.24) together with the known thermodynamic parameters. In this way one obtains an equation for σ, which yields a
realistic value. And also: Applying this ΔT one remains above the shaded area of
Fig. 1.3.

Equation (2.24) is based on thermodynamic considerations, which do not permit
a too large supercooling. However, it is well known that one can find homogeneous
nucleation at temperatures much farther below the equilibrium melting point. But
one can look at Eq. (2.24) as an equation, which only holds, if the character of long
chain molecules is ignored. In fact, the obstructions against association, as occurring with macromolecules, are quite large. Mutually fitting conformations at the
points of contact always depend on a series of interdependent rotations around
adjacent chemical bonds, which are mostly hindered by the lack of space. But these
difficulties promote the shift of homogeneous nucleation to lower temperatures. On
the other hand, obtained junctions will be stable. This fact is guaranteed by the large

distance to the equilibrium melting temperature. This expectation has proved to be true. Experience has shown that nuclei are stable as soon as they are formed at temperatures below the temperature indicated by Eq. (2.24). Only with increasing temperatures, which approach the equilibrium melting point, one observes the relaxation of nuclei. This fact can be observed for shear induced nuclei on Fig. 2.20. A more detailed discussion will be given in the next section.

There remains only one question: What will be the reason for the fact that the number density of stable nuclei further increases so much with decreasing temperatures, if already the nuclei, which are found at higher temperatures, are stable. In fact, the pertinent reason seems to be that the probability for the occurrence of clicks increases as a consequence of decreasing demands for a definitive association. One should not forget that spherulites melt at much lower temperatures then thread-like aggregates: see Fig. 2.20. Also the transition to other crystallization modifications should not be overlooked. In fact, why is the growth speed of spherulites of the β-modification so much larger than the growth speed of the more stable α-modification? (See Fig. 2.11.)

2.3.2.3 The Role of Local Alignments

There is still another point, which did not get any attention in the discussion, which had been carried on so far on nucleation. This point concerns the local alignment of macromolecules. In rheology Pechhold et al. [43] realized the importance of this phenomenon. However, as the three dimensional description was impossible, these authors proposed a two dimensional model. They called it the meander model. However, two dimensions are by far too restrictive. In fact, if a random flight path is drawn in two dimensions on a sheet of paper, a neighboring molecule has to follow a path of almost the same shape, if a high density of lines is required. In this way the authors arrived at their meanders. But also a qualitative consideration of the three dimensional case can be conclusive. In fact, if one selects an arbitrary macromolecule in a quiescent melt, this macromolecule has a statistical shape. But if one now considers a neighboring molecule, one finds that this molecule cannot have an independent arbitrary shape. At least over some part of its contour it must snuggle to the first mentioned molecule. At other parts of its contour it will snuggle to a third and a forth molecule. In fact, if one prepares a series of stiff wire models of statistical shapes and throws them on a heap, one will never get the high density of a melt. The achieved density will be more that of a gas. The conclusion can only be that the molecules must show a sufficient flexibility, so that they can adjust their shapes like cooked spaghetti. And it will also be clear that there will be alignments of widely differing qualities.

In Chap. 1 (Sect. 1.1) a T,S-diagram has been introduced. In this diagram two transitions are shown (Fig. 1.3). The upper horizontal line indicates the melting temperature of ideal crystals. The level of this line gives the equilibrium melting point. The lower double line gives the range of melting temperatures of spherulitic structures. Above this latter range of temperatures one finds a shaded area, where still stable nuclei have been found. One type of those nuclei are the mentioned nuclei, which are obtained by self-nucleation. However, these nuclei are not the only ones, which are found to be stable above the melting temperature of the spherulites. In stressing this fact, a picture is introduced here in Fig. 2.20, which actually must be discussed in Chap. 3, where the flow induced crystallization will be treated. However, for a valid argumentation this picture is transferred to the present section. It shows the relaxation times of shear induced thread-like precursors, which were obtained in a melt of PP at the temperatures quoted at the abscissa. The applied shear rates (in duct flow) were of the order of 1200 s^{-1}, with shearing times of $\approx$0.5 s. The data, which are used in this graph, have a long history. Already in 1989 a pertinent paper was published by Eder, Janeschitz-Kriegl and Krobath [44]. In our first review [15] the quantitative data, as obtained by H.Wippel in his engineer's thesis, were published. Finally, a recollective paper was recently published by Janeschitz-Kriegl and Eder [45], where the above figure was presented. One notices that the relaxation times tend to go to infinity at temperatures much higher than the temperature, where the spherulites melt. An important point was that the precursors, which apparently were thread-like and produced striped oriented structures after quenching, were so tiny that they had no measurable influence on the flow birefringence, which prevailed during the flow. For the determination of the relaxation times various waiting times were introduced between the cessation of the flow and the quench. With increasing waiting times the oriented structures faded according to a single relaxation time. These facts were already shown in the first paper mentioned. The conclusion is that also these nuclei are stable at temperatures well above the melting temperature of the spherulites.

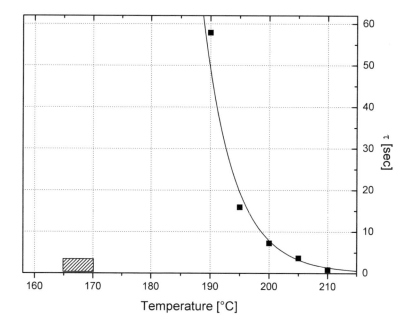

Fig. 2.20 Relaxation times of thread-like precursors versus temperature of shearing at a shear rate of 1200 s^{-1} and subsequent tempering, as valid for an industrial PP [45]. Shaded block on the temperature axis: range of temperatures, where spherulites melt. Courtesy of Taylor & Francis

Interestingly enough, the stability of nuclei does not necessarily mean that they are automatically active centers for growth. Starting with the basic idea of local alignments one must conclude that these nuclei look like fringe micelles. In fact, local alignments cannot be longer than the participating molecules. They will mostly be much shorter. As a consequence one will find so-called tangling ends of molecules on both ends of the regular body of the micelle. Length and thickness of the body of the micelle will be a formal criterion for its orderliness.

If a micelle is very thin, the surface tensions at its side surfaces and, in particular, also on its ends will be very small (according to Tolman and Larson). The latter fact is the consequence of the fact that the tangling ends of molecules have a great freedom in conformation. However, if the conditions for the growth of a lamella are given (see below), this conformational freedom is reduced during the lateral association of other molecules. As the space, which the tangling ends of molecules need in a direction parallel with the growing lamella, is larger than the space required by those parts of the molecules, which are in the body, an effective surface tension is built up during such a growth. The tangling ends get into a squeeze. But this process hampers the growth and will bring it to a halt, if the announced conditions for growth are not sufficiently fulfilled. This is shown schematically in Fig. 2.21.

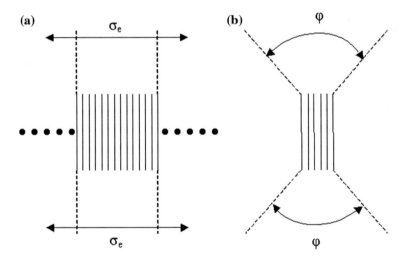

Fig. 2.21 A schematic presentation of the structure (**a**) of a lamella of considerable lateral dimension and (**b**) of a bundle of strands of a lateral dimension insufficient for an effective surface tension σ_e. The freedom of the tangling ends is symbolized for the latter configuration by the cone angle ϕ, according to Janeschitz-Kriegl [46]. Courtesy of Springer Verlag

Fig. 2.22 The course of the
effective surface tension σ_e,
characteristic for the (rough)
surface of a lamella, as a
function of the number N of
laterally associated
macromolecules. At a critical
number N_c the increase of σ_e
will be most conspicuous
[46]. Courtesy of Springer
Verlag

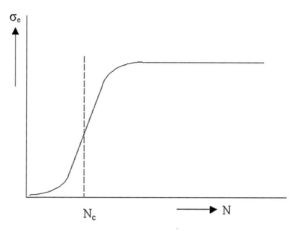

Because of the apparent importance of this conclusion, also a schematic picture
of the dependence of the effective surface tension on the number of associated
macromolecules is given in Fig. 2.22.

As a next point the condition for growth must be formulated. As already
announced, growth can only happen, when the finally obtained lamella is stable.
For this purpose the surface tension in its rough surfaces, where also back-folding
can occur in reducing the tension [47], must be borne in mind. In fact, the positive
free energy per unit surface at the end of a growing fringe micelle must be in
balance with the negative free energy (difference with the fluid), as counted per unit
surface at an internal cross-section of the micelle. But, the negative value of the
latter free energy depends on the length of the body of the micelle, i.e. on its
orderliness. But this means that for a longer micelle the barrier is lower than for a
shorter micelle. And there is also the influence of temperature. Whereas the surface
tension is not very sensitive to temperature (see the role of kT in rubber elasticity),
the internal free energy difference is very sensitive to temperature. (At the equilibrium melting point the latter is zero, whereas the surface tension remains finite!
See Janeschitz-Kriegl [42].) But this means that the negative value of the body free
energy difference increases seriously with decreasing temperature, so that the
barrier decreases rapidly also for those micelles, which are not so well structured.
This fact is reflected for PP by the left side of Fig. 1.1, where the number density of
activated nuclei increases by about five decades, if the temperature is lowered from
about 130 to 85 °C. We are confident that this result also shows the statistical nature
of the local alignments. Certainly there must be many more badly structured
alignments than well structured ones.

These results lead us to the conviction that the local alignments, which are
described above, are transformed into stable athermal nuclei, when the melt is
cooled to a temperature below the temperature range, where the spherulites melt. In
this connection it is also instructive to follow the succession of events happening
during the melting of a spherulite. During this process not only single molecules
will be detached. Certainly also whole bundles of molecules will be separated from

each other. If these bundles become slim enough, they will be stabilized according to the principles outlined above. In our opinion this is a valid explanation for the self-nucleation effect.

Unexpectedly we received a very well-come support for our ideas by Kornfield and collaborators [48] quite recently. These authors used an ingenious novel technique of levitating a small electrically charged particle of about 1 ng in an electric field. If such a particle changes its weight by—say—absorption of the vapor of a fluid, the increased weight can accurately be determined by adjusting the electric field in order to keep the particle at its original height. The method is called "scanning activity gravimetric analysis SAGA". For the above mentioned purpose the authors exposed a tiny particle of poly(ethylene oxide) PEO to a humidity cyclically varying with time. As PEO is water soluble, the humidity (the activity of water) can take over the role of the temperature. It appeared that at a humidity somewhat higher than the humidity, at which the particle dissolved and formed a clear spherical droplet, nuclei survived in the droplet for hours. The formation of the clear droplet was documented with the aid of the proper light scattering (Mie scattering) and by the increased weight. In fact, when the applied maximum humidity was reduced again, crystallization started at a considerably higher humidity than in a droplet, which had been exposed to a much higher maximum humidity, where the nuclei apparently could not survive. Crystallization was detected by the upcoming turbidity. A survival of hours indicated that there was no kinetic effect at stake. A theoretical interpretation was given in terms of the behavior of a special disk-like lamella.

It remains to be emphasized that the tremendous difference between the equilibrium melting point and the melting temperature of the spherulites, as shown in Fig. 1.3, must be a consequence of the melting point depression caused by the surface tension on the lamellae. And Strobl [49] gave a nice description of the pertinent process. It must also be pointed to the fact that ideal crystals cannot be obtained in a direct way. They have always been obtained in a re-crystallization process under high pressure and at temperatures close to the equilibrium melting point, as Wunderlich and also Bassett have shown [50–52]. And this equilibrium melting point has always been obtained in keen extrapolation processes by Hoffman and Weeks and by Marand et al. [53, 54]. These facts are not surprising. In the temperature interval between the melting temperature of the spherulites and the equilibrium melting point sporadic nuclei must have a stretched conformation for the formation of ideal crystals, in which the molecules are also stretched. However, in the melt one will not find any molecules, which are completely stretched in a spontaneous conformation. This also explains, why in this temperature range crystallization kinetics are not observable. Only flow can change this situation, as will be shown in Chap. 3.

2.3.3 Winter's Gel Point

Some time ago Winter and Chambon [55] published a fundamental paper on the exact determination of the gel point (of the time elapsing until gelation) of a cross-linking polymer. For their experiments the authors used a polydimethylsiloxane (PDMS). This polymer could be cross-linked in its fluid state at the ends of the molecules with the aid of tetrasilane. In dynamic mechanical measurements the authors found that at a certain stage of the process storage and loss moduli became congruent over a wide range of frequencies. At too early a stage the loss modulus was larger than the storage modulus. After gelation the storage modulus was the larger one. The criterion for the gel point was:

$$\tan(\delta_c) = \frac{G_c''}{G_c'} = \tan\frac{n\pi}{2} = \text{const.,} \qquad (2.25)$$

where δ_c is the loss angle at the gel point and the shear loss and storage moduli, as obtained at this point, are given in the usual notation. And n is a positive constant equal or larger than one half. In order to exploit this criterion, a frequency sweep over at least five decades is required. Of particular interest is the time t_c, which elapses until this gel point is reached. However, for a practical execution this time must be much longer than the time required for the frequency sweep, which largely depends on the lowest frequency required. In fact, within this sweep the properties of the sample should not change to any extent.

Later Pogodina and Winter [56] used this criterion also for the gelation of i-PP, as caused by incipient crystallization. The authors found that this time decreased remarkably with decreasing temperature, i.e. with increasing undercooling. However, only low degrees of undercooling were permitted. In fact, at larger undercoolings the crystallization process became too fast. Acierno and Grizzuti [57] extended these type of measurements on i-PP down to 135 °C. In order to cope with the faster process, they invented the "inverse"quench. For the purpose the sample was kept for some time at the low temperature. The obtained crystallinity was then stabilized by a quick increase of the temperature to 158 °C, where kinetics were so slow that an accelerated frequency sweep, making use of the spectrum of harmonics, could be carried out. If the right time was chosen for the development at the lower temperature, the criterion of Eq. (2.25) was met at 158 °C. The obtained results are shown in Fig. 2.23. One notices that the gel times are extremely large for our terms. Even at the lowest temperature of 135 °C ($\Delta T = 56$ K) t_c was almost one hour (3500 s). (ΔT is surprisingly large for 135 °C, because a relatively high value of 191 °C was chosen for the melting temperature of the spherulites.) Also, at the gel point the degree of crystallinity is still very low, of the order of a few percent. So far we are not able to lay a connection to our results except for the fact that the cross-links are practically stable at 158 °C, a temperature not too far below the melting temperature of the spherulites. This fact is in accord with our vision on the stability of nuclei.

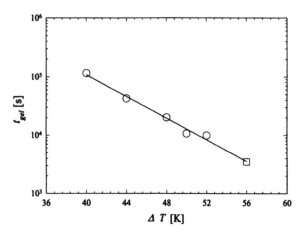

Fig. 2.23 Gel time of i-PP as a function of the degree of undercooling according to ref. [57]. The square symbol was used for the point, which was obtained with the inverse quenching technique. Courtesy of the Society of Rheology

So far it seemed impossible to relate the occurrence of the gel point to our results. This fact puzzled us for a long time. Finally we decided to assume provisionally that gelation started at nuclei, on which also the spherulites grew. This assumption is necessary, because neither of those nuclei can be counted in the quiescent melt at the relevant high temperatures. The reason is their scarceness at those temperatures. However, at least the number of nuclei for the growth of spherulites, as counted at lower temperatures, can be extrapolated to the higher temperatures, where the gelation times can be obtained with the aid of dynamic mechanical measurements. This procedure is described in a paper by Janeschitz-Kriegl and Ewa Ratajski [58]. In this work the areas, which contain the gel-like structure, are assumed to start like clouds around the corresponding nuclei. Fortunately, the results of the said extrapolation of the nucleation density are uncritical. The order of magnitude suffices.

This fact is obvious, when the growth speed of the said clouds is obtained with the aid of Kolmogoroff's Eq. (1.10). In this form this equation is valid for the isothermal case, when all nuclei are born simultaneously. As the gelation becomes manifest only after so long times, the process of cooling to the experimental temperature can be considered as instant. One can also assume that the space is filled sufficiently well with the said clouds, if the argument in the exponential of this equation has become equal to −1. When looking at this equation one notices that only the third root of the numbers per unit volume is needed. The letters G, N and t stand for growth speed, density of nuclei and gelation time.

One can now compare the obtained growth speeds of the clouds with the growth speeds of spherulites in the same temperature range. This comparison is shown in Fig. 2.24 for it-polypropylene. It is of interest that in the range of temperatures above 135 °C the growth speeds of the clouds of the gel are larger than the growth speeds of the spherulites. But the growth speeds of the gel clouds increase more slowly with decreasing temperature than the growth speeds of the spherulites. As a consequence, both growth speeds become equal near 135 °C. This fact means that

Fig. 2.24 Double logarithmic plot ot the growth speed G_g of jellied volume elements against the growth speed G_s of spherulites of ordinary i-PP for a series of crystallizarion temperatures [58, Fig. 4]

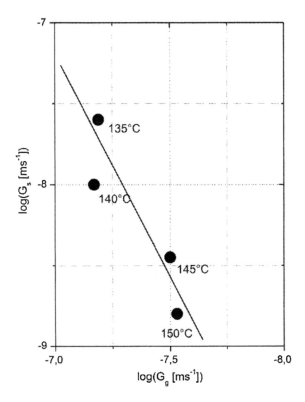

below 135 °C one can no longer decide with the aid of dynamic mechanical measurements, whether gelation is due to some network forming or to spherulite growth.

Because of the uncertainties, which were caused by the extrapolation and by the reading on graphs, we were happy to find an unexpected support with the aid of measurements, which were carried out in the course of the research on flow induced crystallization. As these results will be discussed in the next chapter, it will be sufficient to mention that a birefringent gel, which is induced by shear flow, expands in the quieted down melt at 145 °C much faster than the spherulites.

References

1. Van Krevelen DW (1978) Crystallinity of polymers and the means to influence the crystallization process. Chimia 32:279–294
2. Fisher JC, Hollomon JH, Turnbull D (1948) Nucleation. J Appl Phys 19:775–784
3. Janeschitz-Kriegl H, Ratajski E, Wippel H (1999) The physics of athermal nuclei in polymer crystallization. Colloid Polym Sci 277:217–226
4. Braun J, Pillichshammer D, Eder G, Janeschitz-Kriegl H (2003) Industrial solidification processes in polybutene-1. Part I—Quiescent melts. Polym Eng Sci 43:180–187

5. Stadlbauer M, Eder G, Janeschitz-Kriegl H (2001) Crystallization kinetics of two aliphatic polyketones. Polymer 42:3809–3816
6. Chew S, Griffiths JR, Stachurski JH (1989) The crystallization kinetics of polyethylene under isothermal and non-isothermal conditions. Polymer 30:874–881
7. Janeschitz-Kriegl H, Eder G, Stadlbauer M, Ratajski E (2005) A thermodynamic frame for the kinetics of polymer crystallization under processing conditions. Monatshefte für Chemie (Chemical Monthly) 136:1119–1137
8. Eder G, Janeschitz-Kriegl H (1997) Processing of polymers: crystallization. Mat Sci Tech (VCH-Wiley) 18:269–342
9. Gandica A, Magill JH (1972) A universal relationship for the crystallization kinetics of polymer materials. Polymer 13:595–596
10. Ratajski E, Janeschitz-Kriegl H (1996) How to determine high growth speeds in polymer crystallization. Colloid Polym Sci 274:938–951
11. Turner-Jones A, Aizlewood JM, Beckett DR (1964) Crystalline forms of isotactic polypropylene. Makromol Chem 74:134–158
12. Wunderlich B (1973) Macromolecular physics, vol 1. Academic Press, p 282
13. Lovinger AJ, Chua JO, Gryte CC (1977) Studies of the α and β forms of isotactic polypropylene by crystallization in a temperature gradient. J Polym Sci Polym Phys Ed 15:641–656
14. Janeschitz-Kriegl H, Wimberger-Friedl R, Krobath G, Liedauer S (1987) On the development of layer structures in injected plastic parts (in German). Kautschuk + Gummi. Kunststoffe 40:301–307
15. Eder G, Janeschitz-Kriegl H, Liedauer S (1990) Crystallization processes in quiescent and moving polymer melts under heat transfer conditions. Prog Polym Sci 15:627–714
16. Berger J, Schneider W (1986) A zone model of rate controlled solidification. Plast Rubber Process Appl 6:127–133
17. Stein RS, Rhodes MB (1960) Photographic light scattering by polyethylene films. J Appl Polym Sci 31:1873–1884
18. Keijzers AEM (1967) Light scattering by crystalline polystyrene and polypropylene. Doctoral thesis Delft, The Netherlands, p 23
19. Keijzers AEM, Van Aartsen JJ, Prins W (1968) Light scattering by crystalline polystyrene and polypropylene. J Am Chem Soc 90:3107–3113
20. Van Antwerpen F (1971) Kinetics of crystallization phenomena of spherulites in poly (ethylene terephthalate). Doctoral thesis Delft, The Netherlands
21. Van Antwerpen F, Van Krevelen DW (1972) Light scattering method for investigation of the kinetics of crystallization of spherulites. J Polym Sci Polym Phys ed 10:2409–2421
22. Van Krevelen DW (1990) Properties of polymers, 3rd ed. Elsevier, p 594–603
23. Magill JH (1967) Crystallization of poly(tetra methyl-p-silphenylene)siloxane J Polym Sci A-2, 5:89–99
24. Heijboer J (1968) Study of the movements of cycloalkyl side groups in polymethacrylates by dynamic mechanical measurements. J Polym Sci C 16:3413–3422
25. Ziabicki A, Alfonso GC (1994) Memory effects in isothermal crystallization I. Colloid Polym Sci 272:1027–1042
26. Mandelkern L, Kim H (1968) Temperature dependence of the bulk crystallization rate of polymers. J Polym Sci A-2, 6:965–706
27. Hoffman JD (1964) Theoretical aspects of polymer crystallization with chain folding: bulk polymers. SPE Trans 4:315–362
28. Williams ML, Landel RF, Ferry JD (1955) Temperature dependence of relaxation mechanisms in amorphous polymers and other glass forming liquids. J Am Chem Soc 77:3701–3707
29. Hoffman JD, Davis GT, Lauritzen JI Jr (1976) In: Hannay NB (ed) Treatise on solid state chemistry, vol 3, chap 7. Plenum Press, New York
30. Clark EJ, Hoffman JD (1984) Regime III crystallization in polypropylene. Macromolecules 17:878–885

31. Van Antwerpen F, Van Krevelen DW (1972) Influence of crystallization temperature, molar weight and additives on the crystallization kinetics of poly (ethylene terephthalate). J Polym Sci Polym Phys Ed 10:2423–2435
32. Mandelkern L (2004) Crystallization of polmyers. In: Kinetics and mechanisms, 2nd ed, vol 2. Cambridge University Press, pp 1–204
33. Hoffman JD, Miller RL (1997) Kinetics of crystallization from the melt and chain folding in polyethylene fractions revisited: theory and experiment. Polymer 38:3151–3212
34. Lauritzen JI Jr, Hoffman JD (1960) Theory of formation of polymer crystals with folded chains in dilute solution. J Res Nat Bur Stand 64A:73–102
35. Hoffman JD, Lauritzen JI (1961) Crystallization of bulk polymers with chain folding: theory of growth of lamellar spherulites. J Res Nat Bur Stand 65A:297–336
36. Blundell DJ, Keller A, Kovacs AJ (1966) A new self-nucleation phenomenon and its application to the growing of polymer crystals from solution. Polym Lett 4:481–486
37. Larson MA, Garside J (1986) Solute clustering in supersaturated solutions. Chem Eng Sci 41:1285–1289
38. Larson MA, Garside J (1986) Solute clustering and surface tension. J Cryst Growth 76:88–92
39. Becker R, Döring W (1935) Kinetic treatment of nucleation in supersaturated vapour (in German) Ann Phys 5,24:719–752
40. Tolman RC (1949) The effect of droplet size on the surface tension. J Chem Phys 17:331–337
41. Rusli IT, Larson MA (1987) Solute cluster formation in saturated solutions. In: Stratman, Klein, Melis (eds) Cryst Precip Proc Int Symp. Pergamon Press, p 71
42. Janeschitz-Kriegl H (1997) Conditions of nucleation in crystallizable polymers, a reconnaisance of positions. Colloid Polym Sci 275:1121–1135
43. Pechhold W, Hauber MET, Liska E (1973) Meander model of amorphous polymers. Kolloid Z Z Polym 251:818–828
44. Eder G, Janeschitz-Kriegl H, Krobath G (1989) Shear induced crystallization, a relaxation phenomenon in polymer melts. Progr Colloid Polym Sci 80:1–7
45. Janeschitz-Kriegl H, Eder G (2007) Shear induced crystallization, a relaxation phenomenon in polymer melts. A recollection. J Macromol Sci B Phys 46:1–11
46. Janeschitz-Kriegl H (2003) How to understand nucleation in crystallizing polymer melts under real processing conditions. Colloid Polym Sci 281:1157–1171
47. Keller A (1957) Single crystals in polymers: evidence of folded chain configuration. Phil Mag 2:1171–1175
48. Olsen AP, Flagan RC, Kornfield JA (2006) Manipulation of athermal nuclei in aqueous poly (ethylene oxide) by scanning activity gravimetric analysis. Macromolecules 39:8419–8427
49. Strobl G (1996) The Physics of polymers. Springer, p 160
50. Prime RB, Wunderlich B, Melillo L (1969) Extended chain crystals V. Thermal analysis and electron microscopy of the melting process in polyethylene. J Polym Sci A-2, 7:2091–2099
51. Bassett DC (1981) Principles of polymer morphology. Cambridge University Press, p 168
52. Woodward AE (1989) Atlas of polymer morphology. Hanser Verlag pp 106–115
53. Hoffman JD, Weeks JJ (1962) Melting process and equilibrium melting temperature of poly (chloro trifluoro ethylene). J Res Nat Bur Stand A66:13–28
54. Marand H, Xu J, Srinivas S (1998) Determination of the equilibrium melting temperature of polymer crystals: linear and non-linear Hoffman-Weeks extrapolation. Macromolecules 31:8219–8229
55. Winter HH, Chambon F (1986) Analysis of linear viscoelasticity of a crosslinking polymer at the gel point. J Rheol 30:367–382
56. Pogodina NV, Winter HH (1998) Polypropylene crystallization as a physical gelation process. Macromolecules 31:8164–8172
57. Acierno S, Grizzuti N (2003) Measurement of the rheological behavior of a crystallizing polymer by the "inverse quenching" technique. J Rheol 47:569–576
58. Janeschitz-Kriegl H, Ratajski E (2014) Flow induced crystallization in polymer melts: how Winter's gelation concept fits into the picture. Polym Bull 71:1197–1118

Chapter 3
Flow Induced Processes Causing Oriented Crystallization

3.1 Preamble

The investigations in this field started around the year 1970. From the beginning shear flow [1] and extensional flow [2] experiments were carried out. Another classification was with respect to experiments, where flow was continued until the viscosity of the melt started to increase rapidly [1], and those where the progress of crystallization was followed during flow in a more subtle way by dilatometry [3], by scattering experiments [4] or by a count of upcoming nuclei [5]. In fact, flow can increase the speed of crystallization enormously. Last but not least the pioneering work by Van der Vegt and Smit [6] should not be forgotten. With decreasing temperatures these authors observed an increasing obstruction at the entrance to a capillary, which was used for viscometry. The authors correctly interpreted this obstruction as the result of a crystallization starting under the influence of the extensional entrance flow.

Yet another classification has been with respect to the type of rheometers used. Parallel plate rotational rheometers were often used [3, 4], lately also in [7, 8] (using the apparatus of Linkam Scientific Instruments with glass plates). For high shear rates rectilinear flow is required. In rotational viscometers secondary flow can be very disturbing at high angular velocities. For a suitable rectilinear flow a sandwich construction has been used, where one glass plate moves at a constant distance with respect to another glass plate [1, 9–11]. Recently, an apparatus was constructed by Janeschitz-Kriegl, Ratajski and Eder [12]. In this apparatus the curvature of the flow lines was minimized. In contrast to the mentioned sandwich practically unlimited shear could be realized in this machine. For a discussion of results see later in this script. A rectilinear flow is also obtained, if a glass fiber is drawn through the undercooled melt [13]. Near the surface of the fiber rather high rates of shear can be obtained.

© Springer International Publishing AG 2018
H. Janeschitz-Kriegl, *Crystallization Modalities in Polymer Melt Processing*,
https://doi.org/10.1007/978-3-319-77317-9_3

Also duct flow is of the type of rectilinear flow. Actually in injection molding one has such a situation, if the cavity is a duct of rectangular cross-section of large aspect ratio [14, 15]. However, with injection molding big complications are caused by two facts: the hot polymer melt is injected into a cold mold and, at the beginning, this mold is empty. So, one has a heat transfer problem and the complicated flow pattern at the flow front (the so-called "fountain flow" [16]).

Nevertheless, duct flow experiments can be very useful. A prerequisite is that one starts with an already filled duct, which is cooled down with its content (the sample) from a temperature well above the melting point to the desired crystallization temperature. When this temperature is reached, injection is resumed. No fountain flow occurs in a filled duct. No heat transfer problem exists for the period of flow. Short term shearing is desirable [17]. Of course, the duct must be open at its end. Close to this end windows are placed for an optical observation. It turns out that premature crystallization, which, in principle, seems possible in the still quiescent melt during cooling, is a minor problem. In fact, at not too low temperatures, where flow has already a large effect, kinetics are still extremely sluggish in quiescent melts. These experiments led to the first quantitative interpretation. For the purpose a research group at the CALTECH built a miniature equipment for small samples [18].

3.2 Some Comments of Considerable Reach

As is well-known, the entropy of a rubbery network structure is reduced, if the network is stretched, which means that its free energy is lowered. As a consequence its melting point must increase by stretching [19, 20]. The question is, how far such an effect will also be of importance for flowing polymer melts, in which the macromolecules are stretched and oriented as well. However, any theory describing such an effect is necessarily based on a molecular model. Aware of this difficulty the present author discovered that one can actually measure the change of the free energy [21]. In fact, every experienced rheologist knows that the first normal stress difference N_1, as occurring in steady shear flow of a rubber-like liquid, represents twice the stored free energy. Being purely entropic according to all types of molecular theories, this normal stress leads to the equation:

$$N_1 = 2\Delta G = 2T\Delta S_o, \tag{3.1}$$

where subscript "o" stands for orientation. In fact, the said molecular theories always contain extra assumptions in addition to that of the entropic nature of the forces. If the equations

$$T_0 = \frac{\Delta H}{\Delta S - \Delta S_0} \quad \text{and} \quad T_m = \frac{\Delta H}{\Delta S}$$

are combined, one obtains:

$$T_0 = T_m \left(1 + \frac{N_1}{2\,\Delta H}\right) \tag{3.2}$$

Naturally, the free energy stored during shear flow is always much smaller than the heat of fusion ΔH of a phase transition. Moreover, N_1 can be measured. The most convenient way is the indirect way with the aid of the flow birefringence Δn. One has [22]:

$$N_1 = \frac{\Delta n}{C} \cos(2\chi) = \Delta\sigma\,\cos(2\chi), \tag{3.3}$$

where C is the stress optical coefficient (as known for many polymers), χ is the extinction angle and $\Delta\sigma$ is the principle stress difference. The latter two parameters hold for the given shear rate. This equation holds for the (1,2)-plane, where direction "1" is the flow direction and direction "2" is the direction of the velocity gradient. The light beam is in the "3" direction in this case. A simpler equation is obtained, if the light beam is in the "2" direction, which means that the birefringence is observed through windows positioned in the large side walls of a rectangular duct (of large aspect ratio). One has in this case of a "slit rheometer":

$$\Delta n = n_{11} - n_{33} \approx C N_1 \tag{3.4}$$

In this equation Δn is an average value (half the value at the duet surface) over the cross-section of the duct.

The use of the constant stress-optical coefficient C justifies also the use of Eq. (3.1). In fact, the stress-optical coefficient appears to be constant (at constant temperature) and independent of the molar mass for many polymer melts over three to four decades in the shear rate. In this range the non-Newtonian viscosity decreases for some polymer samples by more than one decade (cf. [22], pp.108–119). But this means that, surprisingly, stress in flow birefringence is clearly dominated by the orientation of chain segments as in rubber. In fact, there is no other reason for—say—stress induced birefringence because of the softness of the melt (see [22], p. 184). There is a simple proportionality factor between the stress and the refractive index tensors of any chain segment, independent of its temporal length (this length should not be too short, of course). The only condition is that with change of the conformation of the segment its internal equilibration must be fast compared with the changes in contact points with the surrounding. The consequence is that Eq. (3.1), which primarily holds for rubber, also holds for flowing polymer melts. In this way, Eq. (3.1) is actually underpinned mainly by an experimental evidence. Probably, one can short-circuit this consideration just by assuming:

$$\frac{\Delta n}{C} \approx 2T\Delta S_0 \tag{3.4a}$$

For our estimate, however, this does not matter. From measurements, as carried out at the time by Wales and Philippoff [23] (see also p. 146 of Ref. [22]), one learnt that for the melt of a LDPE the birefringence was about 10^{-4} at a shear rate of one hundred reciprocal seconds and a temperature of 190 °C. From this result it followed with Eq. (3.4) and $C = 2.1 \times 10^{-9}$ Pa^{-1} that N_1 was about 4.76×10^{-2} MPa. However, for ΔH (per cubic meter!) of LDPE a much higher value of about 130 MPa is calculated from the value of ΔH given in the Polymer Handbook [24]. It is now only necessary to transform the first normal stress difference from 190 to 110 °C. For the purpose the shift factor for this polymer is required. It is found on p. 37 of the said book [22], as taken from the work of Münstedt and Laun [25]. From 190 to 110 °C a shift over about one decade is required. To illustrate this procedure, Fig. 3.1 is introduced. This figure is from the said work of Wales and Philippoff (see also Wales [26]). Of interest for the present purpose is only the upper line on the right side. It holds for $n_{11}-n_{33}$ (filled triangles), as obtained with a slit rheometer. (Open squares are from an extrapolation of cone and plate measurements). If this line is shifted by one decade to the left, one obtains a vertical distance with the original line corresponding to a factor three. So one obtains a N_1

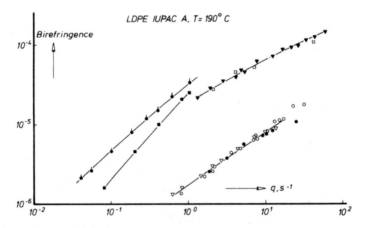

Fig. 3.1 Various types of flow birefringence of a sample of LDPE, as plotted against the shear rate for a temperature of 190 °C [22, 23], Courtesy of Dietrich Steinkopf Verlag. Upper left curves: cone and plate measurements of Δn and $n_{11}-n_{22}$. Upper right curve: n_1-n_{33} from measurements in a slit rheometer. Lower curve: $n_{33}-n_{22}$ from axial capillary measurements, of no relevance in the present case

of about 0.14 MPa for the temperature of the flowing melt of the LDPE. This N_1 is less than one thousandth of the heat of fusion. With a melting temperature of roughly 400 K this means an increase of this melting temperature about 0.1 K. At decreasing rates of shear this value decreases in accord with Fig. 3.1. With HDPE the heat of fusion is much higher than with LDPE. But the elasticity of the melts of HDPE is not much larger. The conclusion is that under usual experimental conditions an increase of the melting point will not be of importance.

This conclusion also holds for polymers with an extra broad distribution of molar masses. It is true that in such a polymer the molecules of the highest molar mass carry most of the load and contribute most strongly to the orientation and the corresponding birefringence of the sample. As Brochard and De Gennes showed [27], the longest molecules, if taken together, would form a new phase of lower free energy, which should separate from the mother phase. However, the highly oriented molecules are dispersed amongst the large number of short molecules. In a relaxation experiment, the orientation of the short molecule will relax faster, but the still highly oriented long molecules will not get together within their prolonged relaxation because the required self diffusion in a direction perpendicular to the previous direction of the stretch will certainly take more time. In fact, only direct neighboring molecules can participate in the nucleation of a new phase.

This conclusion must lead to another comment. As is well-known, with shearing and with extensional flow thread-like nuclei ("shishs") are formed. After the cessation of the flow overgrowth happens on these threads at a much lower rate. In this way so-called "kebabs" are formed. During the flow, however, these shishs can reach considerable lengths well hundred fold the length of single macromolecules. But some authors think that these threads can be formed by the spontaneous association of stretched molecules. However, such a process is unthinkable. Those molecules cannot tumble, as there is no space for tumbling (see the work of Prof. Kornfield's group, which will be discussed below). Probably, still unbound neighboring chain molecules will be rolled out along already existing oriented bundles of molecules.

But it is not only this principal consideration, which is in favor of the idea that long threads can be formed by such a mechanism. In a great number of duct flow experiments, as carried out in Linz and characterized in Graz [17, 28], it has been shown that such a mechanism offers the only possible explanation for the great total shish length per unit volume. This fact will be elucidated later in this chapter.

In this connection the a priori existence of local alignments can be of importance. But nobody knows exactly the mechanism of the internal reorganization, which is necessary for a permanent click. In this respect the reader is reminded of the fact that nuclei are stable, if they are formed at a temperature, which is sufficiently far below the equilibrium melting point. Twenty degrees below the equilibrium melting point will suffice (see Fig. 2.20). It will also be evident that small sized nuclei must not be point-like. During short term flow their length is increased, but has not yet reached a length comparable with their mutual distance. This means that there will still be enough space for the growth of spherulites, which happens as soon as flow is stopped (see the effects of short term shearing against long term spherulite growth).

As has been pointed out in Sect. 2.3.2.3, local alignments must have the shape of fringe micelles. Apparently, these fringe micelles are oriented by the action of the flow, before they grow in the flow direction and become effective thread-like nuclei also at the higher temperature of shearing. Figure 1.1 is suggestive in this respect. In fact, after flow at higher temperatures the number density reaches a level as in the quiescent melt at lower temperatures.

A similar improvement of the fringe micelles is also suggested by the effect of pressurization. By the application of about 100 bar the number density of effective nuclei considerably increases in iPP, as two exchange students (E. Rexhepaj and R. Vulkicevic) have experienced during their (short) stays in Linz. These results are reproduced in Fig. 3.2. One must assume that the fringes are ironed by the pressure.

The influence of pressure is also emphasized in a paper by Ma et al. [29]. In this paper the purpose of pressurization is to increase the effective melting point and, as a consequence, the degree of supercooling. In this way a fast quench can be replaced, which cannot easily be realized in a cooling process. In fact, the pressure can be changed much faster in a sample than the temperature. For the change of the temperature one needs a complicated cooling cycle (see Sect. 2.2.1), which cannot be applied in a more complicated apparatus, which is developed for a pretentious purpose, say for flow induced crystallization. In this respect the reader was reminded by the said authors of the thermodynamics, as formulated by Clapeyron. With a polyethylene an increase of the effective supercooling by 8 K could be reached within 25 s, when the pressure was increased to 300 bar.

Quite old work points to another growth mechanism of fringe micelles. Boon et al. [30] and Van Krevelen [31, 32] reported of the following experiment with a melt of isotactic polystyrene: This melt was first cooled down to the glass transition

Fig. 3.2 Number densities of nuclei in melts of iPP against pressure for various temperatures, unpublished work (see the text)

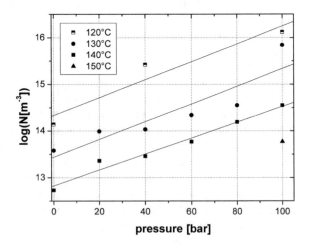

temperature, where it apparently did not crystallize. Then it was reheated to various higher temperatures for crystallization. The authors found that the number densities of spherulites became in this way by factors of one hundred thousand higher then number densities, which were obtained, when the samples were directly cooled from the molten state to the said crystallization temperatures. Apparently, many local alignments were developed at the low temperatures so that they could become effective at higher temperatures. This is shown in Fig. 3.3. Also here no orientation effect was at stake. The large effect, as shown in Fig. 3.3 for the described detour in cooling, is comparable with the large effect of shearing, as shown on the right side of Fig. 1.1.

The authors are tempted to ascribe another badly understood effect to a similar mechanism. As reported be Keller and Kolnaar [33], shish-kebabs are found in stretched network structures. During stretching the shishs appear almost immediately. But these shishs, which are observable under the electron microscope, do certainly not consist of single stretched molecules. Close neighbors must have been

Fig. 3.3 Number densities versus reduced temperature in iPS. Lower line is found after a direct cooling to the temperature of crystallization, upper line stands for the effect of an intermediary quench to the glass transition temperature [31]. On the lower curve one also finds some points for iPP and PET. Courtesy of Brunner Verlag, Zürich

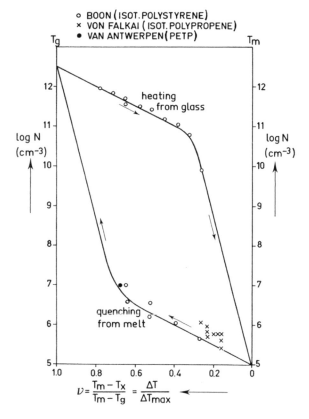

aligned. The overgrowth (kebab formation), however, which occurs, when the quick stretching is finished, takes a lot of time.

A further comment fits into this discussion. In one of the above paragraphs short term shearing was already mentioned in passing. In fact, the technique of short term shearing (or stretching) has been introduced by us in Ref. [17]. It turned out that it enables us to separate nucleation during short time shearing in an elegant way from the growth process occurring slowly afterwards. (See the mentioned differences in the time scales at those high temperatures). The criticism given in Mandelkern's book [34] is not to the point. Mandelkern claims that "chain relaxation dynamics will play an important role in crystallization under these conditions". It is interesting that quite a number of authors succumb to this misconception. Apparently, none of these authors has ever carried out the necessary experiments. In fact, at Linz we were very early with those experiments. The title of our first paper on this subject of 1989 was: "Shear induced crystallization, a relaxation phenomenon in polymer melts" [35], where the rate of disintegration of associated macromolecules is described. In our first review of 1990 this concept was worked out more precisely [21]. This relaxation phenomenon is characterized by a single relaxation time, which increases with decreasing temperatures much faster than the relaxation times of free or entangled molecules (zero shear viscosity). In this connection one should look at Fig. 2.20 of the previous chapter. From this figure one can learn that in the majority of experiments, which were carried out below the melting temperature of the spherulites, the shishs were stable immediately after their formation. We already knew this fact, when we started the experiments, which led us to the results reported in Ref. [17]. But this means that "chain relaxation dynamics" do not play any role in these experiments. As has been indicated in previous sections of this monograph, also small ("short") nuclei can be considered as stable at temperatures below the melting temperature of the spherulites. There are also strong indications that after shearing the increased number of seemingly point-like nuclei remains constant over long times. In all these cases we apparently have to do with aggregates of a stability surpassing by far the stability of the state of orientation of loose macromolecules. Many details will be treated in the subsequent sections.

In particular, it will become evident that, so far, no valid molecular theory does exist for the said growth processes. In fact, as long as for the quiescent melt the topology of the local alignments remains unexplored, no useful insight can be expected into the much more complex processes occurring under the conditions of flow. Admittedly, a beginning has been made with a treatment of the interaction between neighboring molecules by the introduction of the reptation model [36, 37] (from wriggling of snakes). However, as only the average influence of the surroundings is treated in this model, one ends up with the behavior of the singled out molecule. This approach suffices for the conditions of flow, where neighboring molecules are exchanged frequently. But such an exchange does no longer happen, when the melt is quieted down. This remark seems disillusioning. However, it is better to face up to the reality than to trust some approaches, which are based on insufficient insight. In this connection it must be emphasized, that the orientation of the molecules is a necessary but not a sufficient condition.

3.3 Survey of Activities in the Field of Flow Induced Crystallization

In recent years the increase of the number of publications in this field resembles almost an explosion. The early activities are sketched in the preamble. However, when at Linz University the present author decided to start in this field in 1983, it seemed as if interest had dropped off. Nevertheless, interest was pushed by the problems arising in injection molding. In this connection not only the mold filling problems were of importance. As is well-known, those problems arise when a mold of a more complicated shape has to be filled. In this respect untimely solidification in narrow passages can play an undesired role. Under those circumstances low filling rates are required, when fast filling rates are desirable. In fact, low filling rates are also in contrast to the requirements of heat transfer. However, we know that high shear rates and also local extensional flow can shift the temperature of solidification to much higher values. From these experiences it became evident that some quantitative experiments would be helpful.

As already mentioned, two main streams have been recognized so far. In Linz we concentrated on the structure development. In this connection not only duct flow experiments should be mentioned. A very simple experiment consists in drawing a glass fiber through the undercooled melt. The processes occurring on the surface of the moving glass fiber can be observed under the microscope, if windows of glass are incorporated. Other investigators used rotational viscometers, which contain transparent parallel plates of glass or plates with Beryllium windows for simultaneous X-ray measurements. In the second main stream the samples were kept flowing until the viscosity of the undercooled melts showed an upswing. At first sight one would be inclined to treat these main streams separately and compare the results in a final discussion.

However, many experiments were of a mixed nature. Also the understanding and the planning of the experiments depended strongly on the insights obtained up to those moments of decision. So, in the following the historic development will be followed, even if, as a consequence, sometimes complicated situations have to be unraveled. We start with the description of duct flow experiments, as carried out in Linz in the late eighties and in the early nineties. We believe that results obtained with these measurements, namely the first successful phenomenological growth model and the extraordinary relaxation behavior of thread-like precursors, have been decisive for a further development of the field.

3.3.1   Duct Flow Experiments

3.3.1.1   Short Term Shearing ("Step Strain") Experiments [17]

For these measurements an ordinary single screw extruder has been adapted. A schematic presentation of the pertinent arrangement is given in Fig. 3.4. There are two essential parts. One of them is the lengthy duct of rectangular cross-section of large aspect ratio. This duct furnishes a practically two dimensional flow pattern, as the influence of the small side walls can be disregarded in the middle of the flow field. Near the end of this duct a pair of glass windows is placed in the opposite large duct walls. Near the entrance a pressure gauge is mounted flush in one of the large side walls. At the entrance a valve is placed. The second essential part is the accumulator. It is mounted on a reservoir, which connects the extruder with the duct. This accumulator is a tube with a honed inner surface, allowing a sliding fit to the plunger inside. This plunger is loaded over a lever with an exchangeable weight. The choice of this weight determines the pressure in the melt. At the lower end of the accumulator there is a second valve, which serves as a bypass.

With the plunger in a low position the duct is filled in an ordinary extrusion process. During this process the machine is kept as a whole at a temperature well above the equilibrium melting point of the polymer. The required heating is achieved by electric heating bands. Such a heating band is also wound around the cylindrical outer surface of the part containing the duct. This part consists of two halves enclosing the duct. In a second step the valve at the entrance to the duct is closed and, after a proper relaxation time, which is relatively short for the high processing temperature, the duct is cooled with the aid of a heat transfer fluid, which is pumped through the bodies of the said two halves. In this way the chosen

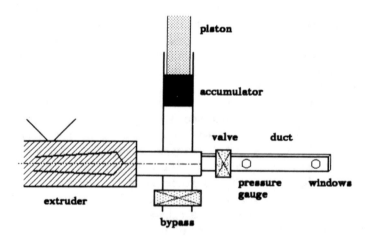

Fig. 3.4 Schematic presentation of the arrangement for the short term shearing experiment [17]. Courtesy of Hanser Verlag. Picture kindly provided by Prof. M Cakmak, Akron, when the author stayed there

lower temperature of the crystallization is reached with a melt, which does not remember previous maltreatment. At the same time the extruder is switched on again in order to push the plunger up in the accumulator. The experiment proper is started, when the valve at the duct entrance is opened. In this way hot melt of the original temperature is pushed into the cooler duct under the constant entrance pressure guaranteed by the weight on the lever. However, before this hotter fluid reaches the windows, flow must be stopped suddenly. This is accomplished by opening the bypass. Of course, also shorter periods of flow were possible. In these experiments only fluid of the proper crystallization temperature passes the windows. The pressure near the windows remains always low (no pressurization!). The high temperature at the entrance is required mainly in avoiding crystallization in the converging (extensional) entrance flow [6]. After a proper waiting time the duct was quenched to the temperature of tap water and the solidified sample was extracted from the duct. This was a very convenient method for iPP, as the solidified samples of this polymer did not stick to the walls.

Two types of measurements were carried out. One type of measurement, using the windows, was concerned with the course of the optical properties (birefringence, depolarization). The other measurements were carried out on cross-sections of the solidified sample (optical and electron microscopy). In Fig. 3.5 the said optical measurements are demonstrated for an industrial PP with a molar mass $M_w = 289\ 000$ [21]. The chosen parameters were: $T_c = 150$ °C, shear rate at the wall: $84\ s^{-1}$, shearing time 4 s. Distance from the entrance to the center of the windows was 115 mm, cross-section of the duct was 10×1 mm. The box-like pressure profile is demonstrated by the dashed line. The flow birefringence is shown together with this pressure profile. It disappears quickly and almost completely at the end of the pressure treatment. The optical retardation, as caused apparently by the insetting crystallization, is given by the wavy curve. The total transmitted intensity, as normalized by its initial value, decreases continuously with time because of the upcoming turbidity. At 14 s there are kinks in both curves because of a change in the speed of the recorder. The time, when the first peak reaches its

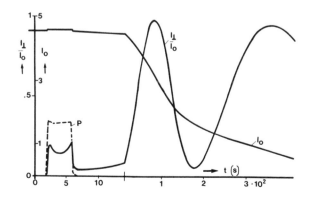

Fig. 3.5 Reduced light intensities versus time, as obtained with crossed polars for a short term shearing experiment on the undercooled melt of an industrial PP, as described in the text [21] Courtesy of Pergamon Press

Fig. 3.6 Continuously monitored optical retardations as functions of the monitoring time, as obtained with an industrial PP after shear treatments of 108 s^{-1} at the duct wall and a temperature of 150 °C. Shearing times in seconds are indicated at the respective curves [17]. Courtesy of Hanser Verlag

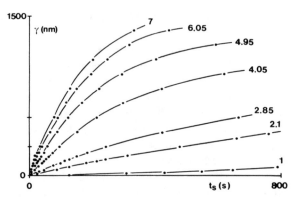

maximum, is taken as a measure for the progress of the crystallization. This time gives the moment, when a retardation of half a wave length is reached.

In Fig. 3.6 the continuously measured optical retardations are given as functions of time for a variety of previous flow conditions, as applied to an industrial PP. Temperature (150 °C) and shear rate (108 s^{-1}) were kept constant in this series of experiments. Only the shearing times were varied. They are given in seconds at the ends of the curves. The optical retardations were determined with the aid of the well-known equation:

$$I = I_0 \sin^2\left(\frac{\pi \Gamma}{\lambda}\right), \qquad (3.5)$$

where Γ is the retardation, λ is the wave length in the medium, I is the observed light intensity, as obtained with crossed polars, and I_0 is the transmitted total intensity, which is weakened by the turbidity. One notices that even the longest shearing time of 7 s is only 5% of the corresponding waiting times. For the other shearing times one obtains much lower percentages. The shearing times are of a much lower order than the waiting times.

For illustration a micro-photograph, as taken with crossed polars from a cross-section of a solidified sample, is shown in Fig. 3.7. This cross-section was parallel to the previous flow direction and perpendicular to the large duct walls. The conditions of shearing were: a temperature of 150 °C, a shearing time of 7 s and a shear rate of 72 s^{-1} at the wall. Near the wall a highly oriented zone is shown on this figure. The picture covers a distance of 350 μm from the duct wall. This picture resembles cross-sections taken from injection molded samples. However, the flow conditions were strictly isothermal and there were no influences of a flow front.

Fig. 3.7 Cross-section
through a solidified sample of
an industrial PP, as obtained
after shearing at 150 °C for
7 s at a wall shear rate of
72 s^{-1} [17]. Courtesy of
Hanser Verlag

The times, at which a retardation of half the wave length (of 550 μm) was
reached, were plotted against the shearing times on double logarithmic scales for
three temperatures: 143, 150 and 157 °C. In Fig. 3.8 the results for 150 °C are
reproduced. The steep negative slopes are remarkable. A provisional theoretical
consideration suggested the preparation of a set of straight lines on a transparent
paper. The slope of these lines was minus two and the vertical distances between
these lines were chosen to be four times the logarithm of the ratio of neighboring
wall shear rates. This grid was put on the plot of the experimental points in a
judgment by the eye. The pertinent wall shear rates are shown near the lines. If the
adjustment is accepted, one arrives at the conclusion that the shear treatment is
governed by the following surprising combination of variables:

$$q_w^4 \, t_s^2, \tag{3.6}$$

where q_w is the shear rate at the wall and t_s is the shearing time. But a look on the
graph suggests that the distance between the lines should even be larger. The
corresponding picture for a temperature of 138 °C has been shown on several
previous occasions. It resembles very much the picture shown in Fig. 3.8. In particular also for 138 °C the vertical distance between the constructed lines is not big
enough.

Fig. 3.8 Double logarithmic plot of the observation time, at which half the wave length of retardation is reached, against the shearing time for the mentioned industrial PP. The belonging shear rates at the duct walls are given near the lines. The temperature of experimentation was 150 °C [17]. Courtesy of Hanser Verlag

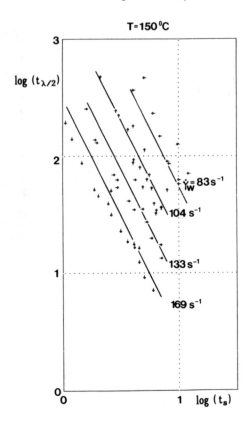

3.3.1.2   A Preliminary Interpretation of the Results of Section 3.3.1.1

One of the basic principles of such an interpretation is that the process cannot depend on the direction of shearing. But this means that the dependence on the shear rate q must be expressed by an even function of this shear rate. We decided for simplicity to use the square of the shear rate as the simplest even function. (In this connection it may be mentioned that q^2 can also be considered as the second invariant of the rate of deformation tensor). And the forth power of the shear rate, as suggested by the just mentioned experimental results, can only be obtained by the superposition of two processes, which both depend on the square of the shear rate. For the purpose we assumed that in the first process primary nuclei are created spontaneously, whereas in the second process thread-like precursors grow on those primary nuclei. As a consequence the first differential equation reads:

$$\frac{dN}{dt} = \left(\frac{q}{q_{a,n}}\right)^2 g_n - \frac{N}{\tau_n}, \qquad (3.7)$$

where N is a number of nuclei per unit volume, $q_{a,n}$ is a shear rate of activation, g_n is a factor with the dimension $s^{-1}m^{-3}$ and τ_n is a relaxation time. With an initial condition of $N(0) = 0$ one obtains by an integration of Eq. (3.7):

$$N = \left(\frac{q}{q_{a,n}}\right)^2 g_n \tau_n \left[1 - \exp\left(-\frac{t_s}{\tau_n}\right)\right].$$ (3.8)

If one assumes that the growth of thread-like precursors starts immediately, when the primary nucleus is born, one can put τ_n equal to infinity and obtains:

$$N = g_n \left(\frac{q}{q_{a,n}}\right)^2 t_s.$$ (3.9)

The total length L_{tot} of thread-like precursors per unit volume becomes in this way:

$$L_{tot}(t_s) = 2 \int_0^{t_s} \frac{dN}{ds} L(t_s - s) ds.$$ (3.10)

The differential equation for the growth of the threads reads now:

$$\frac{dL}{dt} = \left(\frac{q}{q_l}\right)^2 g_l - \frac{L}{\tau_l},$$ (3.11)

where q_l again is characteristic shear rate and g_l is a constant with the dimension ms^{-1}. As we already know from Fig. 2.20, for iPP the relaxation time τ_l will be of importance only close to the equilibrium melting point. At temperatures below 190 °C it will be practically infinitely large. This certainly holds for measurements below 160 °C. By integration of Eqs. (3.10) and (3.11) and initial condition $L(0) = 0$ one obtains:

$$L_{tot}(t_s) = 2 \frac{q^4}{q_{a,n} q_l} g_n g_l \tau_l \left[(t_s - \tau_l) + \tau_l \exp\left(-\frac{t_s}{\tau_l}\right)\right].$$ (3.12)

Previously this equation was used by us for a more detailed consideration [38]. However, this approach turned out to be unrealistic. If the exponential function is developed in terms of t_s/τ_l, its first term cancels against the first term in square brackets and one obtains a dominating term:

$$L_{tot} = g_n g_l \frac{q^4}{q_{a,n}^2 q_l^2} t_s^2.$$ (3.13)

This equation furnishes exactly the product given in Eq. (3.6).

When Liedauer obtained his results, there was a dilemma with respect to an adequate interpretation. Certainly, the argument that one would need an even function in the shear rate, was correct. But, why should it be the most simple even function, the square of the shear rate? In fact, only recently, more than twenty years later, an answer for this question was found. An interpretation in terms of convection has become more and more probable. In fact, in shear flow neighboring layers of the fluid (of a thickness comparable with the dimensions of macromolecular coils) move with somewhat different speeds. But, apparently, these relative movements cause more effective touches between molecules than the Brownian motion, which is responsible for encounters in the quiescent melt. (See the interpretation of Fig. 1.1). The new idea is now, to split the square q^2 of the shear rate into $q.q$ and multiply the second factor q with the shearing time t in obtaining $q.t = \gamma$, which is the total shear strain. The remaining factor q can be interpreted as a measure for the frequency of successful encounters. Nevertheless, we know that the probability for an encounter, which leads to a successful "click", remains rather low. But this fact does not matter for our new interpretation. One can imagine that during the course of the deformation there occur regular impacts with neighbor, which can lead to lasting contacts. We also know that those contacts become more fragile, if the equilibrium melting point is approached (see Fig. 2.20). Interestingly enough, by this interpretation the simplest function q^2 is justified.

One can now understand, why the non-Newtonian flow has no direct influence. It must be used only for the calculation of the shear rate. In fact, those shear rates varied considerably in the experiments by Liedauer (see Fig. 3.8), where one would actually expect a direct influence of the non-Newtonian flow. But the absence of such a direct influence is a strong allusion to the role of the convection. Also the small influence of the temperature seems to point in the same direction. A more detailed interpretation will be given later, where as a consequence of prolonged shearing at low shear rates and high temperatures loose structures are formed first. In fact, those loose structures can be deformed easily in a see of moving but still unbound macromolecules.

Postponing further deliberations for later one can say that the given quite rough interpretation of the experimental results leads to a quite useful description of the facts. Within the frame of this description one can say that there is only one adjustable parameter, which is given by

$$\frac{g_n g_l}{q_{a,n}^2 q_l^2} = f(T, MWD). \tag{3.13a}$$

where the abbreviation MWD means molar mass distribution. It can also be said that the temperature dependence of this parameter is rather small [17]. This fact becomes evident, as soon as the influence of the rate of lateral growth is eliminated from the results of Fig. 3.8 and other figures of this type. Only this growth rate, which—after cessation of flow—is taken to be equal to that in a quiescent melt, is well known to be temperature dependent (see Fig. 2.15).

3.3.1.3   Optical and Electron Microscopy [17]

With respect to the optical measurements the first puzzle is the linear increase of the retardation at short waiting times (see Fig. 3.6). In fact, if one would assume the growth of cylindrical bodies around the shishs, one should expect a zero initial slope and a quadratic increase with time, because of the quadratic increase of the surface of cylindrical bodies. But this puzzle is solved by a look on cross-sections through a solidified sample, as cut in a direction perpendicular to the previous flow direction.

In Figs. 3.9 and 3.10 two pertinent pictures are shown, as obtained with the aid of an electron microscope after staining with RuO_4. The pertinent flow data, as applied to the used PP, were a wall shear rate of $72\ s^{-1}$, a shearing time of 7 s at a temperature of 150 °C. Figure 3.9 shows a cross section taken close to the duct wall. This picture covers an 11 μm distance from this wall surface. Figure 3.10 covers an area between 55 and 66 μm from the duct wall.

First of all one notices on both pictures that lateral growth occurred in the quieted down melt along spokes starting at the points, where the shishs pierced through the surface of the cut. This fact can explain the mentioned linear growth of the retardations with time. Secondly, one observes that the average mutual distance between the shishs increases with the distance from the wall. For the purpose of a later interpretation it is of importance to stress that both pictures are made in zones of rather high orientation, where no spherulites are formed.

As in the above theoretical formulation the total length of shishs per unit volume is calculated, one needs a relation between this length and the observed mutual distance between the shishs. As a model for this purpose the unit volume of a stack

Fig. 3.9 Electron micrograph of a cross-section perpendicular to the previous flow direction, covering a distance of 11 μm from the large surface of the sample. Data for the sample of PP and conditions of flow are given in the text [17]. Courtesy of Hanser Verlag

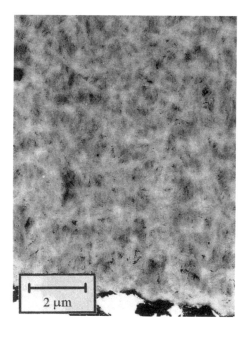

2 μm

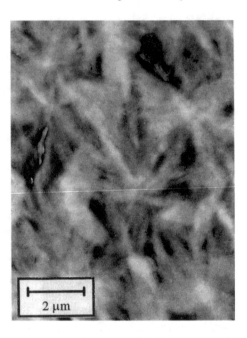

of wooden logs of equal thickness D and of a uniform length of one meter is
considered. An easy calculation gives:

$$L_{tot} = \frac{2}{\sqrt{3}\,D^2} \qquad\qquad (3.14)$$

(By the way: One should not forget that the dimension of L_{tot} in this equation is
correct, because m/m^3 = 1/m^2). Counting is quite difficult on Fig. 3.9. But on
Fig. 3.10 one finds only about 15 points, where shishs pierced through the surface
of the cut. The longer side of the picture corresponds with 11 μm and the shorter
side with 8.5 μm. So one calculates an area of 91×10^{-12} m^2. If one divides this area
by 15, one obtains per shish an area of 6.07×10^{-11} m^2. If this area is put into
Eq. 3.14 instead of D^2, one obtains for L_{tot} an enormous value of 1.9×10^{11} m/m^3.

However, the said enormous value immediately becomes less impressive, if it is
divided by the number of nuclei per unit of volume, which can serve as starting
points for the growth of these lengthy nuclei. In fact, we do not know this number.
Let us take by way of trial a number density of nuclei, as given on the right side of
Fig. 1.1, being about 10^{16} m^{-3}. We have guessed that the transition to thread-like
nuclei would occur at specific works certainly bigger than 7 MPa. If the said L_{tot}
of $\sim 2\times10^{11}$ m/m^3 is divided by 10^{16} m^{-3}, one arrives only at 2×10^{-5} m or
20 mm. Such a short length cannot even be recognized easily on the microscopic
picture. But this means that the number of nuclei, which can serve as starting points
for the growth of thread-like nuclei, must be much lower than 10^{16} m^{-3}.

If for the shear rate profile in the duct the so-called power law is used with
$n = 0.33$, one has:

$$q(x) = q_w \left(\frac{2x}{H} \right)^{1/n} , \qquad (3.15)$$

where x is the distance from the central plain and H is the duct height. Half the duct height is 350 µm. If 55 µm are subtracted, one obtains for distance x from the center 295 µm. Inserting these values into Eq. (3.15), one obtains for the shear rate at 55 µm from the wall $q(x) = 44.8$ s^{-1}, if the shear rate q_w at the wall is 72 s^{-1}, as quoted above. The forth power of the ratio of these shear rates is 6.5. But this means that on the area of Fig. 3.9. a number of $15 \times 6.5 \cong 98$ points should be found. It is difficult, to carry out a count on this figure, but the order of magnitude of the number of points is certainly correct. A better idea is certainly, to make the two pictures coincide by enlarging Fig. 3.9 properly. The corresponding linear magnification of Fig. 3.9 would be $\sqrt{6.5} = 2.56$. Instead of carrying out this enlargement completely, we decided to reduce Fig. 3.10 by a factor 0.54 and enlarge Fig. 3.9 by a factor 1.4. In fact $(1.4/0.54) \cong 2.6$. This procedures were carried out on both figures. For comparison these figures were turned by ninety degrees. So Fig. 3.11 enables us to carry out a convenient comparison. The conclusion can only be a satisfactory one.

Finally one can say that a direct optical comparison confirms the indirect conclusion, which was drawn from the course of the optical retardations, as described in the previous section. The total length of shishs per unit volume is surprisingly high. It supports our idea that shishs are created by a systematic growth process and

Fig. 3.11 Comparison of the pictures from Figs. 3.9 and 3.10 after proper change of magnifications [17]. Courtesy of Hanser Verlag

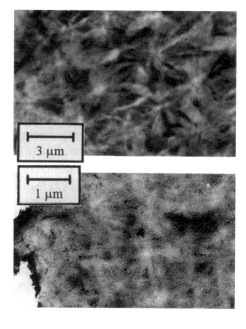

not by the sporadic association of a few perfectly stretched molecules [33], which are particularly long and loosely distributed in space.

Another interesting insight into the nature of the "shishs" is provided by prof. Kornfield in collaboration with a group of Japanese authors [39]. These authors applied small angle neutron scattering to three special samples of iPP. These samples were mixtures of three fractions of different molar masses. In each sample one of the fractions was Deuterium labeled. Of particular interest was the sample, in which the low molar mass fraction was Deuterium labeled. After selected re-melting the highly oriented surface layer, as obtained after duct flow, the authors found that the same amount of the said low molar mass fraction was present in the surface layer and in the bulk. This means that no selection of the high molar mass molecules has occurred for the surface layer.

3.3.1.4   More About Structures and a Growth Mechanism

Some years ago the research group at the CALTECH developed a miniature apparatus for short term shearing in duct flow [18]. The purpose of this machine is that small quantities of specially prepared samples can be investigated. In Fig. 3.12

Fig. 3.12 Schematic presentation of the shear unit [40]. A description is given in the text. Courtesy of Elsevier

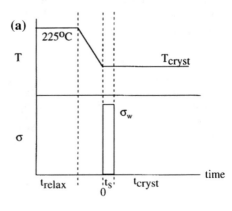

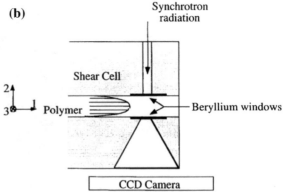

a cross-section through this machine is given. The upper part gives a protocol of the working of the machine, if iPP is to be investigated. After what has been said so far, no comment is necessary. The lower part is sometimes equipped with Beryllium windows situated near the end of the duct [40]. These windows are necessary for WAXD measurements. The same holds for the conical aperture at the exit of the beam. It enables the recording of a wide angle fiber diagram. The dimensions of the duct are: 63.5 mm × 6.35 mm × 0.5 mm. The polymer melt is pushed into the duct by a plunger system driven by gas pressure. Maximum throughput is about 100 mg. The X-ray measurements were carried out at the chosen temperature of crystallization (see the upper part of Fig. 3.12). As an X-ray source a synchrotron is used with a wave length of $\lambda = 130.7$ pm.

In the first paper of a series of papers [41] visible light intensities were measured through crossed and parallel polars. In this respect the reader is asked to look at the previous Fig. 3.5. On this figure one has to observe the course of the optical retardation, which is found during the period of 5 s, when the pressure was applied. In the beginning this retardation shows an overshoot, which relaxes soon. However, a renewed increase of the retardation is noticed, which does not completely relax, when the pressure is taken off. At the early times, no particular attention was paid to this phenomenon. But the research group of Prof. Kornfield made up for this omission. It was observed by this group that the formation of thread-like precursors ("shishs"), which apparently happened during the period of flow, was linked to this renewed upswing of the retardation.

The said authors produced an extremely interesting graph, which is shown in Fig. 3.13. It consists of two parts. On the abscissa of both parts only the time t_s, which elapses during shearing, is plotted. On the ordinate axes the reduced light intensities $I_\perp/I_{tot}$, as obtained through crossed polars, are plotted as in Fig. 3.5. All measurements were carried out at a shearing time of 5 s and at a fixed shear stress of 0.06 MPa. For clarity the curves are vertically offset. Otherwise they would be too close to each other. The curves represent measurements taken at the following temperatures: 135, 140, 145, 150, 155, 160, 165, 170 and 175 °C, starting from the bottom. Only at the highest temperature of 175 °C the birefringence relaxed completely after cessation of the flow. The left figure shows a seemingly absurd result: The higher the temperature was, the faster occurred not only the relaxation of the overshoot but also the growth of the renewed retardation. In the right graph the time axis is re-scaled according to the time-temperature superposition principle [42]. For the purpose the times t_u, at which the lowest points of the curves were obtained, were divided by a temperature dependent shift factor a'_T, so that all minima occurred at the same reduced time t_u/a'_T (see the right part of Fig. 3.13). And the course of this shift factor agreed practically with the course of the shift factor a_T^{WLF} of iPP according to Ref. [42]. This is shown in Fig. 3.14. This result rightly caused the authors to choose "Evidence of a kinetic pathway to nucleation" as a subtitle of their paper.

However, at this point the present authors would like to add a comment. The shear rates at the duct wall, which are not mentioned in the papers from the CALTECH people (because of the uncertainty in their determination), must

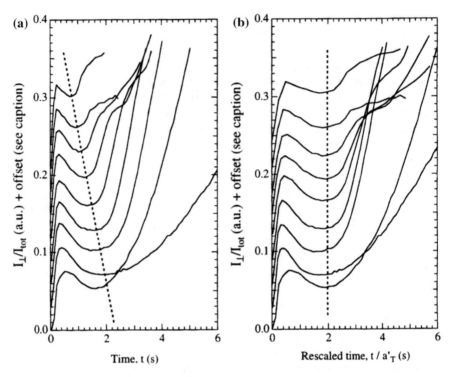

Fig. 3.13 Reduced light intensities through crossed polars, as obtained during shearing of an isotactic PP at a fixed shear stress of 0.06 MPa and a series of temperatures. These temperatures were 135 to 175 °C in intervals of five degrees, starting at the bottom [41]. Courtesy of the American Chemical Society

nevertheless play a qualitative role. These shear rates certainly increase with the experimental temperature, if the shear stress is kept unchanged. But this means that the specific work has also been on the increase. The argumentation is elucidated in Fig. 3.15. For simplicity a two dimensional sketch is prepared. In this picture attention is paid to the behavior of a small fraction of particularly long molecules, which are strongly oriented and highly stretched during shearing. In fact, experience teaches us [33] that those long molecules play a prominent role in the formation of thread-like precursors. With three blobs the centers of two long macromolecules and an aggregate (local alignment) are given. At the center of the aggregate the relative flow profiles are sketched for two shear rates, a smaller one and a larger one. The centers of the two long macromolecules are drawn at equal distances from the center of the aggregate. The corresponding vectors, however, point into very different directions. With the distance between two straight lines, which are drawn parallel to the flow direction on either side of the aggregate, the transverse bounds are indicated of the critical contact region, within which a stretched macromolecule can touch the aggregate, when passing by in the flow direction. Obviously the contact region of the aggregate is most important.

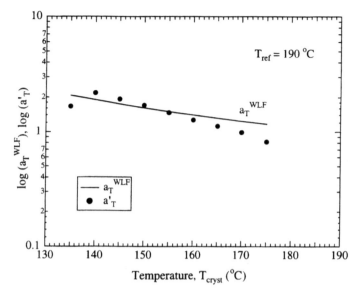

Fig. 3.14 Comparison of the shift factor a'_T for t_u (filled circles) with the shift factor a_T^{WLF} (solid curve) for iPP [41]. Courtesy of the American Chemical Society

Fig. 3.15 A sketch of the circumstances, which are relevant for association of macromolecules to aggregates in a shear field. For an explanation see the text

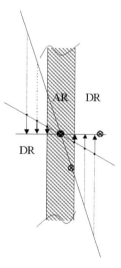

In Fig. 3.15 this region is indicated as the association region AR. The field outside this region AR may be called the disentanglement region DR. In fact, a macromolecule, which passes by with its center somewhere in DR, will never touch the central aggregate. If such a molecule reaches the upper right quadrant of Fig. 3.15, its distance to the said aggregate will increase without bounds to infinity. In fact, it seems quite realistic if one assumes that in the average all molecules continue to move parallel to the flow lines. Entanglements will tug on all sides, before they break. Complete disentanglement will be the consequence. The higher the shear rate is, the more partners will be brought up in AR during a certain time span.

However, as has been stressed previously, most touches will remain without consequences also inside the area AR. But this fact also means that the strict boundaries of AR must be softened. An additional complication rises only, if the experiments are carried out within a temperature range close to the equilibrium melting point, as bordered by Eq. 2.25. In this range the obtained junctions are only temporary. They can be saved only by a sudden cooling process. Nevertheless it should be emphasized that experiments in this high temperature range can be eye opening. This fact will be shown in the section on the relaxation of nuclei and in the section, where results of uninterrupted flow will be discussed. This comment seems necessary in view of the fact that most investigators have avoided to enter this range of rather high temperatures. In fact, a big contrast can be expected between results, which are obtained with it-PP at about 140 °C and a shear rate near 100 s^{-1} or with the same polymer at 190 °C (or higher) and a shear rate near 1 s^{-1}. The reader will see that one can learn a lot from the latter type of experiments. Figure 2.20 should provide a foretaste at least with respect to the choice of temperatures.

Returning to the experiments at the CALTECH further important results must be mentioned [43, 44]. The authors carefully prepared mixtures of two polypropylenes of extremely different molar mass distributions. For one case these distributions are shown in Fig. 3.16. The sample of the lower average molar mass formed the basic polymer. The high molar mass sample of particularly narrow molar mass distribution was added in small quantities. It was prepared by fractionation. The samples were first dissolved in xylene at 141 °C, where the mixing occurred. The mixtures were precipitated in methanol and dried. So, no mechanical degradation of the high molar mass molecules occurred during the mixing. Against oxidation the samples were protected by an antioxidant.

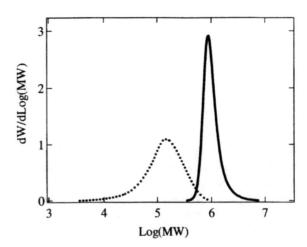

Fig. 3.16 Molecular mass distributions of the fraction of iPP [44]. Courtesy of the American Chemical Society

For the high molar mass fraction the overlap concentration of the coils was calculated according to De Gennes [45]. For this fraction an overlap concentration of 7×10^{-3} g/cm^3 was calculated. In Fig. 3.17 the reduced light intensities $I_\perp/I_{tot}$, as obtained during flow and, in continuation, also after cessation of flow, are given as functions of time for various concentrations of the high molar mass fraction. These concentrations were in g/cm^3 $\times 10^{-3}$: A: 0.00, B: 1.76. C: 3.51, D: 7.00, D: 13.9. The shear stress applied was always 0.14 MPa and the temperature of the experiments was 137 °C. Near the curves one finds the shearing times in seconds and the throughputs in milligrams. In parts A, B and C one notices almost no influence of the increasing number of long molecules. In part D, however, at the critical concentration of the overlap, one notices a considerable shift of the maxima to shorter times. This is true at least, if the throughput is equal or higher than 0.75 mg. For twice the overlap concentration (part E) this effect is even more pronounced. The heights of the maxima were also greater, which should point to an increased transparency. The authors emphasize that a sharp increase of the crystallization speed, as found at the overlap concentration, points to a cooperative effect of the molecules: The orientation of singled out molecules does not explain this behavior. This statement is very satisfying for the point of view of the present author. The fact that the lowest throughputs do not follow this trend, may be explained by a too small specific work. This should point to the building up of a threshold, as the present author would see it in terms of the specific work.

The authors of Ref. [44] also found another very interesting fact: The undiluted high molar mass polymer did not produce a highly oriented zone near the duct surface. In this respect this high molar mass polymer behaved like the low molar mass polymer. The explanation was that, in contrast to the mixtures, the load was not concentrated on the few long molecules but homogeneously distributed over all molecules, so that the degree of orientation of single molecules remained low.

In Ref. [40] a WAXD study is presented. The result, which seems most impressive to the present author, is shown in Fig. 3.18. A sample of iPP with a weight average molar mass M_w of 300.000 g/mol and an M_w/M_n of 6–8 was investigated. The melt of this polymer was subjected to short term duct flow. As a high energy X-ray synchrotron source was used (of a wave length of 130.7 pm), a fiber diagram could already be obtained before the end of shearing. The details of this treatment are given in the caption to the figure. Interestingly enough already after a flow time of only 10 s (at a wall shear stress of 0.06 MPa and a temperature of 141 °C) a complete fiber diagram for the α-crystal modification was obtained. After cessation of flow this diagram remained unchanged during a waiting time of as much as 1200 s. As the interior of the sample experienced lower stresses (and much lower shear rates in a velocity profile approaching plug flow—because of the non-Newtonian behavior of the melt!), the crystallization apparently did not progress very much into the interior during these 1200 s at 140 °C. Otherwise a diagram without any indication of orientation would prevail because of the overwhelming thickness of the inner part of the sample. In fact, the sample was quenched after a little more than 1200 s. On a cross-section it was found that the highly oriented surface layer was only very thin. Apparently 1200 s are much too

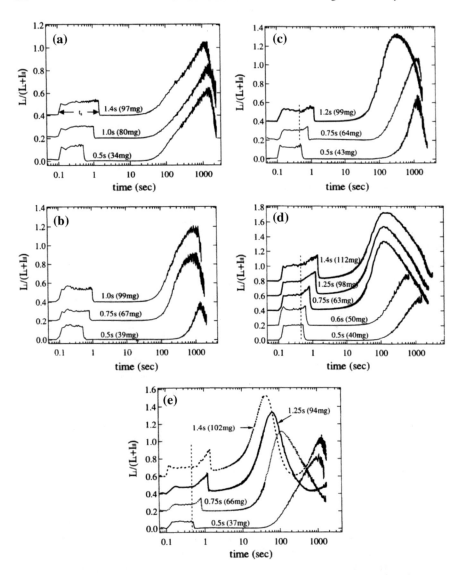

Fig. 3.17 Relative light intensities ($I_\perp/I_{tot}$) through crossed polars during and after short term shearing at a shear stress of 0.14 MPa and at a temperature of 137 °C. These intensities are plotted against time for the samples A through E, which contain an increasing amount of the high molar mass fraction [44]. Observe the offsets in times and intensities. Courtesy of the American Chemical Society

short at 140 °C for a space filling spherulitic crystallization in the practically quiescent part of the melt (see the extremely low nucleation densities and growth speeds, as shown in previous Figs. 1.1 and 1.2 for quiescent melts of iPP near 140 °C).

(a)   **(b)**

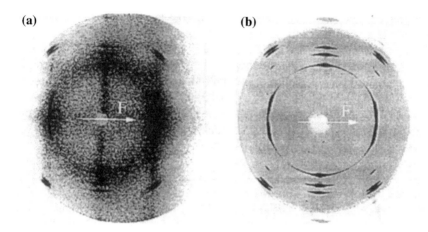

Fig. 3.18 In situ fiber diagrams, as obtained on a sample of iPP, when crystallized under shear (T = 141 °C, σ_w = 0.06 MPa, t_s = 12 s). The flow direction is horizontal. **a** is obtained by acquiring data for 10 s during the period of shearing, **b** is the diffraction pattern obtained after 1200 s during an acquisition time of 40 s. The data have been normalized for the acquisition time [40]. Courtesy of Elsevier

The authors of Refs. [43, 44] arrived at a series of important conclusions:

(a) The formation of thread-like precursors (shishs) is a cooperative process, in which several molecules take part.
(b) Long molecules greatly enhance the growth of thread-like precursors, but only mildly enhance the formation of nuclei, from which threads can grow.
(c) The total length of threads per unit volume increases up to a saturation value.
(d) For concentrations of the high molar mass fraction below the overlap concentration one can observe the finite lengths of single threads with the aid of TEM. Most threads were less than 20 μm long.
(e) The abrupt transition between the skin layer and the spherulitic core provides a measure for the threshold stress of ≈0.12 MPa, which hardly varies with the concentration of long molecules.
(f) The highly oriented skin layer contains also many short molecules. In fact, the concentration of long molecules has been too small in the bulk for filling the space within the skin layer.
(g) Threads do not tumble because of growing continuously. And also, for slender particles the tumbling period increases strongly with their length, as Jeffrey has shown [46].

Probably, a few comments should be added to this list. The transition between the skin layer and the core has not always been so abrupt as in the work of the CALTECH group. Our experience with an industrial polypropylene was that at lower extrusion speeds only a fine grained layer was found near the wall, at medium speeds a highly oriented layer existed close to the wall, more inwards taken over by a fine grained layer, to which the spherulitic core followed in the middle. At sufficiently high extrusion speeds there actually was only an abrupt transition from the highly oriented skin to the spherulitic core. It will not be easy to explain these facts. With respect to the addressed absence of the tumbling one may add that with short term shearing the region close to the wall undergoes something like a planar extension, during which tumbling hardly can be expected.

Duct flow experiments appeared to be particularly useful for the creation of highly oriented structures. As just mentioned, however, sometimes also fine grained structures were found. Our cooperator Liedauer was lucky to take some beautiful photographs. On one of them a clear boundary could be observed between a highly oriented skin layer and a fine grained layer underneath [38]. This picture is reproduced here as Fig. 3.19 for the following flow conditions: $T = 150$ °C, $q_w = 68$ s^{-1}, $\sigma_w = 0.055$ MPa, $t_s = 15.6$ s. Interestingly enough, from the point of view of the shear stress there was only a small gap from $\sigma_w = 0.033$ MPa to $\sigma_w = 0.078$ MPa, within which the transition occurred from an exclusively fine grained surface layer to an exclusively highly oriented skin layer. (For these extreme situations the shear rates at the wall and the shearing times were 19.7 s^{-1}, 39.3 s and 169 s^{-1}, 2.45 s, respectively. In the CALTECH work a critical shear stress of about 0.12 MPa was obtained, which clearly lies above our bounds). The pity with all these experiments is that the applicable shearing time necessarily

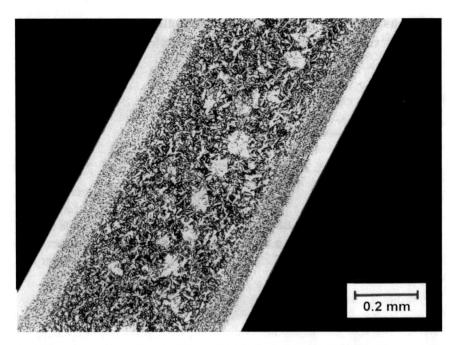

Fig. 3.19 Cross-section through a sample of PP ($M_w = 322000$, $M_n = 47000$) after short term shearing in a flat duct for 15.6 s at a temperature of 150 °C with a shear rate of 68 s^{-1} and a shear stress of 0.0553 MPa, both at the large duct wall [38]. Courtesy of Hanser Verlag

decreases with increasing shear stress and shear rate. As a consequence it becomes very difficult to discern between the exclusive influence of the shear stress and the influence of the integrated specific mechanical work as a whole. In fact, this work increases linearly with the shearing time, which becomes shorter and shorter with increasing shear stress.

In the above paragraph shear stresses and shear rates, as occurring at the large duct wall, where quoted for the PP used in Linz. However, the reader, who wants to check these results, needs the rheological equation of state, which has been used for the purpose. Before giving this equation, which actually holds only for the steady state, one has to explain, why this equation is useful nevertheless. There is a rule of thumb. According to this rule steady state is reached in shear creep experiments after a total shear of about 5 units. But the total shears are much larger in the present case. The pertinent equation, which is ascribed to Carreau [47], reads:

$$\sigma_w = \frac{\eta_0 q}{[1 + (\lambda q)^m]^{\frac{1-n}{m}}} .$$
(3.16)

The data for 150 °C are: $\eta_0 = 21380$ Pa.s, $\lambda = 1.8$ s, $m = 0.537$ and $n = 0.34$.

3.3.2   Flow Induced Small-Sized ("Point-Like") Nuclei

3.3.2.1   Rectilinear Shearing Between Glass-Slides

One of the difficulties with duct flow is the marked change of the shear rate with increasing distance from the duct wall. The shear stress does not change so rapidly with this distance. This fact is due to its linear dependence on the distance from the central plane, which follows from the simple demand that forces must always be in balance. In contrast, the rheological behavior of the melt is reflected by the dependence of the shear rate on the distance from the central plane.

In order to escape this dilemma, the use of a sandwich construction was introduced at Linz University. For the purpose two glass slides are used as confining parallel plates. Both plates have a length of 15 cm and a thickness of 0.2 cm. The width of the lower plate is a little larger, namely 3 cm. A sheet of the polymer ($6 \times 1 \times 0.1$ in centimeters) is sandwiched between these plates and sheared after a proper heating cycle over maximal 5 cm. In this way a maximum shear of 50 units can be obtained. On both sides of the polymer sample Teflon strips of 0.1 cm thickness are inserted. A proper, slightly lubricated weight secures the contact between the polymer and the confining surfaces. The minimization of friction in the driving system is a delicate matter. With the application of high shear stresses one has no problems. The upper plate is set going by a piston, which moves in a pneumatic cylinder connected to an air pressure system. As the sheared surface decreases in the direction of the flow from 6 to 1 cm during the full movement of 5 cm, the force must be reduced for keeping the shear stress constant. In fact, the

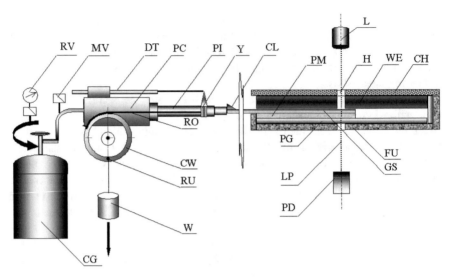

Fig. 3.20 Shear unit, as used in Linz. *PM* polymer melt, *PG* bottom plate of glass, *GS* glass slide, *FU* furnace, *CH* cover with heating elements, *WE* weight, *H* holes, *LP* light path, *L* laser, *PG* photo diode, *CG* steel cylinder for pressurized gas (diminished graphic reproduction), *RV* reduction valve, *MV* magnetic valve, *PI* piston, *DT* Displacement transducer, *CL* clamp, *Y* yoke (perpendicular to the plane of the drawing), *RO* pair of ropes (one in front and the other behind the cylinder), *CW* pair of coupled cart wheels (behind each other), as mounted firmly on their common axle, *RU* rung connecting the wheels, *W* adjustable weight [10]. Courtesy of Springer Verlag

polymer melt behaves more like a fluid than like a rubber. A mechanical device serves for a proper counter force increasing with the distance, over which the upper plate is shifted. This device, which is loaded with a proper weight, actually keeps the shear stress practically constant. The sandwich is embedded in a chamber, which can be heated and cooled properly. Glass plates are preferably used because of their transparency. A light beam can pass through the glasses in every position. So the optical retardation can be determined during flow and afterwards. In Fig. 3.20 a drawing of this machine is presented.

The mentioned mechanism (see the cart wheels, the yoke and the ropes) modifies the force on the upper glass slide in the following way:

$$F_{red} = P.A - M g \sin(\alpha), \qquad (3.17)$$

where P is the gas pressure, A is the area of the piston in the pneumatic cylinder, M is the adjustable mass of the weight, g is the acceleration of gravity and α is the angle, by which the rung is turned away from its lowest starting position. If the radii of the cart wheels is large enough, one has approximately $\sin(\alpha) \approx \alpha$. This means that the required linear dependence of the counter force on the applied shear is readily realized. Smaller total shears than the maximum shear of 50 can be realized,

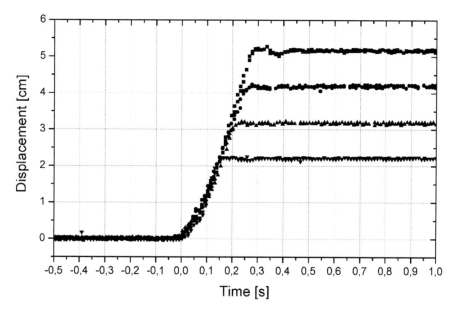

Fig. 3.21 Piston displacements versus shearing times for the melt of an industrial PP at a shear stress of 84 kPa, for increasing total shears between 20 and 50 units and at a temperature of 145 °C. The shear rate common to all curves can be estimated to be 150 s^{-1} [10]. Courtesy of Springer Verlag

if the starting position of the piston is chosen more closely to the position, where the piston is stopped in any case.

The shear stress can readily be determined with the aid of Eq. (3.17). If this shear stress is multiplied by the obtained total shear, one obtains the reduced mechanical work applied. In Fig. 3.21 some results, as obtained with this machine, are shown. One notices the extremely short shearing times, which are realized for total shears of 20, 30, 40 and 50.

Avoiding the influence of friction one can also calculate the applied specific work. For the rate of specific work one has $\eta.q^2$, where q is the shear rate, which can be read from the graph (see Fig. 3.21) and $\eta(T,q)$ is the viscosity of the melt at the temperature and the shear rate of the experiment. Mostly one can rest content by multiplying the mentioned rate of specific work by the time of shearing in order to obtain the total specific work applied. If the shear rate should change too much, one could introduce an integral over the time.

After the shearing process the sample is kept at the temperature of shearing, until total optical depolarization is reached. This depolarization is indicated by the fact that the reduced intensities, which are obtained through parallel and crossed polars, both reach a value of one half. At the same time the total intensity goes nearly to zero. When the shearing time at 145 °C was 0.25 s, the waiting time to total depolarization took about 100 s. This is a remarkable difference. After this stage was reached, the samples were quenched to room temperature and removed from the apparatus. Cross-sections were made parallel and perpendicular to the flow direction (in the 1,2 and in the 2,3 planes). With this machine no highly oriented structures could be achieved. The number of spherulites, which were cut per unit surface, was raised to the power 3/2 for an estimate of the number densities of nuclei, in analogy to the method applied on samples solidified from quiescent melts. In this way the data were obtained, which led to the right side of Fig. 1.1, where the logarithms of the number densities were plotted against the specific work. Remarkably, at relatively high temperatures these number densities reached the same level as the number densities, which were obtained after quenches of quiescent melts at much lower temperatures. By the way, shearing at those low temperatures was not useful because of the badly defined initial conditions. In fact, a complex machine as the one shown in Fig. 3.20, cannot be quenched fast enough.

Before the results, as obtained in these shearing experiments, will be discussed in more detail, a particularly interesting feature of this shearing experiment will be demonstrated in Fig. 3.22 [11]. In this figure two diagrams are shown. The upper graph shows the course of the displacement of the upper plate with time, as occurring on the polymer of Fig. 3.19 with a shear stress of 106 kPa at 150 °C. The total shear of 50 units was reached within 0.23 s. The lower graph shows the changes of the reduced intensities with time through parallel and crossed polars after cessation of flow. It took almost 200 s before any change could be observed. After about 400 s a cross-over occurred. It followed a maximum in the intensity through crossed polars and a corresponding minimum for parallel polars. The measurement finally became quite inaccurate because of the increased turbidity of the sample. Nevertheless, there is no doubt about the fact that intermediately an optical anisotropy was found. This happened regularly with this type of measurements. As no measurable anisotropy was found shortly after cessation of flow, one has to conclude that the birefringence build up was the consequence of lateral overgrowth on tiny oriented thread-like anisotropic particles, which were formed during the short period of flow. And it was also clear that these anisotropic particle acted like "point-like" nuclei giving rise to ordinary spherulites. In fact, on the cross-sections not a bit of anisotropy could be found. Later in this section a more striking example will be given with the aid of a polymer, which produces a much lower number of nuclei. As a consequence the spherulites of this polymer can grow out to a much bigger size permitting the observation of the footprints of their formation.

Fig. 3.22 A special shear treatment of the PP of Fig. 3.19 [11]. The conditions were: a temperature of 150 °C and a shear stress of 106 kPa. In the upper part one finds the displacement of the upper glass plate with time. The lower graph gives the courses of the reduced intensities through parallel and crossed polars, after cessation of flow. Courtesy of Elsevier

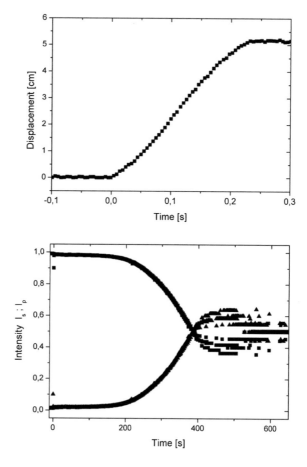

For the moment the results, as obtained with the PP of Fig. 3.19, will be considered more closely. For the purpose the results depicted on the right side of Fig. 1.1 will be plotted on double logarithmic scales. This plot is shown in Fig. 3.23. One immediately notices that the dependence of the number density on the specific work is highly non-linear. For a polybutene-1 Wolkowicz [5] has shown the non-linear dependence of the density of nuclei on the shearing time as early as 1978. Actually, also in the majority of the experiments leading to the present Fig. 3.23 a time dependence is shown. As a matter of fact, a constant shear stress was used for these experiments (84.6 kPa). Only the points at a specific work of about 25 MPa and the points below 0.2 MPa were obtained along different routes. For our PP and a temperature of 160 °C a slope of about four is obtained. For 140 °C this slope is still about three. The lines seem to converge with increasing specific work. They seem to meet at about 25 MPa. This high specific mechanical work had been achieved in duct flow. In this connection one has to look at Fig. 3.19 [38]. The pertinent data are taken from the boundary between the highly oriented layer and the fine grained layer and show, where the transition from a multiplication of nuclei to a creation of a

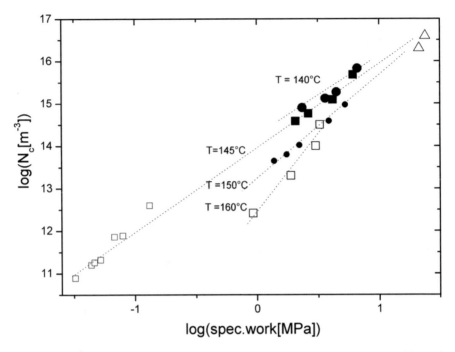

Fig. 3.23 Double logarithmic plot of the number density of nuclei versus the specific work applied on a PP of $M_w = 322000$ and $M_n = 47000$ at several temperatures. Except for the points at specific works below 0.2 MPa, where extensional flow was applied [48, 49], and for the points near 25 MPa, where duct flow was applied [38], all points were obtained at the same shear stress of 84.6 kPa. As a consequence, actually pure time dependences are shown in the middle range [10]. Courtesy of Springer Verlag

highly oriented structure can be expected for this polymer. For very low specific works data from extensional flow are taken [48, 49].

As this subject is of such an importance, also another graph is shown. For the purpose, Fig. 3.24 has been prepared [11]. This figure shows for the same PP the increase of the said number density as a function of the applied specific work, if at 150 °C a great variety of shear stresses is applied. These shear stresses varied from 25 to 250 kPa. Within the expected error margin all points fall on the same line. And also: this line coincides nicely with the line for 150 °C in Fig. 3.23. The reader may notice that the shear stress and the specific work have the same physical dimension, i.e. Pascal. Only, the values of the applied shear stresses are at least one decade lower than those of the specific works.

The extremely important findings, which are described in the last but one paragraph of this section, are supported by results, which have been obtained with another polypropylene. This polypropylene had before undergone degradation with the aid of a peroxide and showed, as a consequence of this treatment, a much lower density of nuclei. But this fact offered more space for an individual investigation of the development of the nuclei. This investigation will be described in Sect. 3.3.2.3.

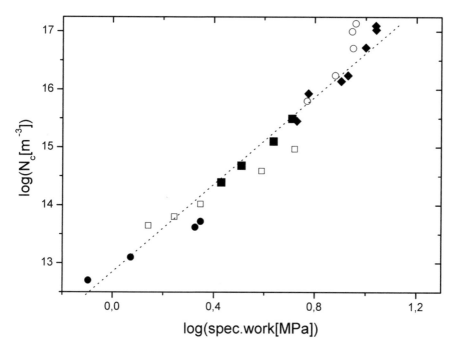

Fig. 3.24 Double logarithmic plot of the number density of nuclei versus specific work for the polymer of Fig. 3.23 at a temperature of 150 °C. A variety of shear stresses has been applied: full circles 25 kPa, full squares 51 kPa, open squares 85 kPa, open circles 127–250 kPa, full diamonds 106–250 kPa [11]. Courtesy of Elsevier

3.3.2.2   Influence of Extensional Flow on the Number Density of Nuclei

Figure 3.23 also shows points below a specific work of 0.2 MPa. As already announced, these points were obtain in extensional flow. A sketch of the apparatus, which has been built for the purpose, is shown in Fig. 3.25   [48]. The main goal was a machine, where the samples had a considerable thickness also after their stretching, so that cross-sections could be prepared and the obtained morphology could be observed. The machine was called the windbix (after the Dutch word "windbuks" for air-gun). In fact, the force, which must adequately decline during the extension for keeping the tensile stress constant, is created by gas pressure in the cylinder shown on the right side of the picture. The pressure in the chamber above the piston is continuously reduced during its downwards movement by the fact that during this movement more and more air can escape through a vertical slit in the pneumatic cylinder. On the left side the cylindrical sample is shown in its initial position inside a heating cylinder. This cylinder is drawn downwards just before stretching is started. The sample is extended in a heated glass cylinder with the aid of a thin wire. This wire connects the sample with the piston in the pneumatic tube. For the purpose it moves over two pulleys. Further details can be found in the caption of the figure.

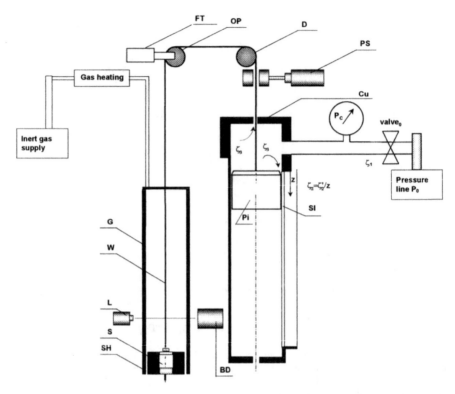

Fig. 3.25 Windbix extensional rheometer. Left side: *SH* sample holder, *S* sample, *L* Laser, *W* wire, *G* glass tube, *OP* overhead pulley, *FT* force transducer, *D* sample length transducer, *PS* pneumatic brake, *Cu* cover, *Pi* piston, *Si* slit, ζ_3 friction coefficient for gas flow through the leaks, $\zeta_2 = \zeta'_2/z$ friction coefficient for gas flow, decreasing with increasing length z of the slit [48]. Courtesy of the Society of Rheology

The notorious necking of the sample was avoided by the use of flexible "clamps". From thin silk fabrics circular specks were cut. In the centers of these specks little ringlets were mounted. These specks were pressed on (glued to) the warmed-up end surfaces of the sample. The wire was tied to the ringlet on the upper side. At the lower side the ringlet was fixed in the center of the bottom plate. When the extension was started, the silk specks folded up like little umbrellas. In this way the circumferences of the end-surfaces were reduced properly.

Two pictures are of particular interest [49]: In Fig. 3.26 the morphologies of two samples of a high molar mass PP are shown. On one sample a very small amount of specific work of 0.01 MPa was applied. On the other sample a ten times higher specific work of 0.1 MPa was applied. These treatments caused very different morphologies. The reader may have his own judgment. With respect to the left picture the number density of nuclei on the right picture is increased by a factor of one hundred. In spite of the application of a stretch no thread-like precursors were

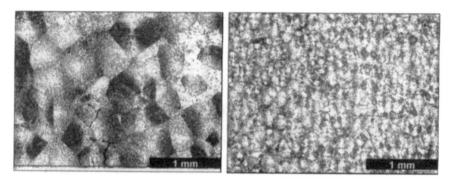

Fig. 3.26 Morphologies after extensional flows for a sample of PP, characterized by $M_w = 492000$ and $M_n = 97000$ [49]. The left picture was obtained at 145 °C after the application of 0.01 MPa, the right picture after 0.1 MPa. Courtesy of the Society of Rheology

Fig. 3.27 Courses of the reduced light intensities, as obtained through parallel and crossed polars after the application of a tensile stress of 10 kPa for 3.5 s at a temperature of 145 °C to the polymer of Fig. 3.26 [49]. Courtesy of the Society of Rheology

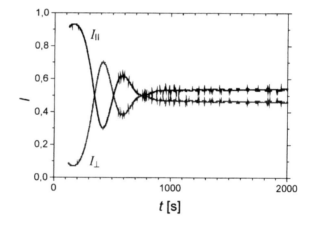

found with these low loads. These pictures resemble very much those, which have been obtained after the application of various specific works in shearing.

In connection with the fact that in the morphologies only spherulites have been found, a reproduction of the course of the optical properties, as monitored after the cessation of the flow, is of particular interest. In Fig. 3.27 the reduced intensities through parallel and crossed polars are shown as functions of the waiting time. To the polymer of Fig. 3.26 a tensile stress of 10 kPa was applied during 3.5 s at 145 °C. Even after this low load a considerable temporary birefringence was registered during the formation of spherulites. Again, one must conclude that tiny anisotropic nuclei, as coming up during the short period of flow, were the basis for the final growth of spherical crystalline entities. One fact is clear in this connection: At the end of the flow treatment the lengths of these nuclei must still be small compared with the mutual distances between these nuclei. Otherwise there would be no space for the growth of spherulites.

3.3.2.3   Results Obtained with a Peroxide Degraded Polypropylene

The present author cannot abstain from presenting a more direct proof for the correctness of the conclusion just drawn. For the purpose, however, a polymer is needed, which under all circumstances contains a much lower number of nuclei than the samples of the polypropylene used so far. With such a polymer the spherulites can grow out to a much bigger size. If by chance a cross-section goes exactly through the heart of such a spherulite, traces of the origin of this spherulite will be visible. It turned out that such a polymer was a peroxide degraded polypropylene. Apparently, the chemical structure is damaged by this treatment to such an extent that the association of molecules becomes much more difficult. Such a polymer was investigated in the Linz laboratory in the course of a routine investigation of a greater number of industrial polypropylenes.

First of all, a three dimensional picture is shown in Fig. 3.28, which is similar to the picture shown in Fig. 1.1. However, when looking at the scales one notices quite large differences in the orders of magnitude. In fact, for the virgin polymer of Fig. 1.1 one finds a number density of nuclei of 10^{16} m^{-3} at 100 °C. For the degraded polymer this density is almost three decades lower.

For this polymer a picture, which has already been published, is shown as Fig. 3.29 [50]. On this figure the effects of short term shearing are shown for a temperature of 150 °C. This figure consists of three parts. On the uppermost part the shear treatment is demonstrated. Shearing took less than 0.1 s for a total shear of 45

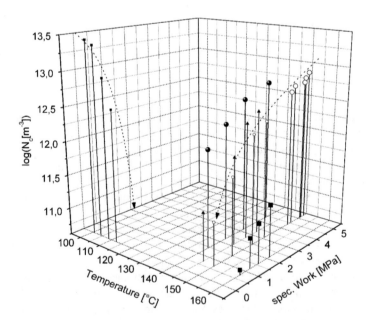

Fig. 3.28 Three dimensional plot of the logarithm of the number densities of nuclei against temperature after fast quenches (left horizontal axis) and specific work (right horizontal axis) for a peroxide degraded PP (M_w = 160000 and $M_w/M_n \approx 2$). Unpublished work by E. Ratajski

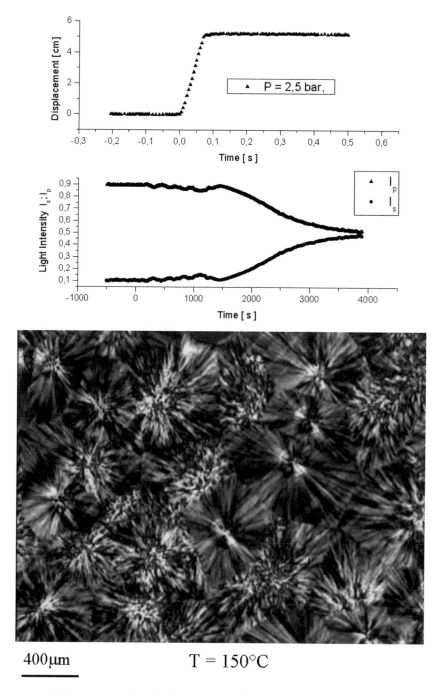

Fig. 3.29 Displacement profile of a short term shearing experiment on the peroxide degraded PP of Fig. 3.28 and the pertinent intensity curves, as obtained through parallel and crossed polars. Also the corresponding morphology of the crystallized sample is shown. Temperature and shear stress were 150 °C and 210 kPa, respectively [50]. Courtesy of Hanser Verlag

units at a shear stress of 210 kPa. On the middle part one can follow the courses of the reduced light intensities, as obtained through crossed polars after the cessation of the flow. One of the axes of this cross has alternatively been under zero and under 45° with the previous flow direction (lower and upper curve). An indication for the occurrence of a crystallization process is found at about 500 s of waiting time (after only 0.1 s shearing time!). After about 1200 s the lower curve shows a slight maximum, when the polarizer is parallel to the duct walls, and a slight minimum, when the polarizer is under 45° with the previous flow direction. As these extrema are not very distinct, one must conclude that only a small portion has been transformed into anisotropic nuclei. But this fact is in agreement with our conception. Complete depolarization was found only after about 4000 s. The lowest picture of Fig. 3.29 shows a cross-section through the solidified sample. The direction of previous shearing is under 45° with the frame of the picture (lower left corner to upper right corner). In the middle of the picture two spherulites can be seen, which obviously were cut almost exactly through their centers. At these centers one notices pairs of little bright and fan-like formations, which are oriented in a direction perpendicular to the previous flow direction. Obviously, these formations have their origin at short thread-like anisotropic nuclei oriented in the flow direction. With more rigorous flow treatments these short nuclei would have grown out into shishs, before the spherulites could have started their growth. In fact, one can observe with this degraded polymer rows of spherulites, which have their origins at clearly visible and seemingly endless strings oriented in the previous flow direction.

The present author has realized later that this figure leads to another remarkable conclusion. Apparently the birefringence, which is observed shortly after the cessation of the flow (see the course of the upper curve on the second picture of Fig. 3.29 at short times), is caused by thin threads of associated macromolecules, which have been formed during the extremely short period of the shear flow. In this respect the rather high shear rate of 650 s^{-1}, which was applied, should be considered. Interestingly enough, these threads appear to be less stable than the spherulites, which become dominant so much later at the unchanged temperature. This conclusion was verified for a normal PP much later, when we had at hand a new machine, which enabled the realization of a homogeneous flow field also for long shearing times (see Janeschitz-Kriegl et al. [12], see a later section).

Quite recently a paper, which clearly supports the view presented here, has been detected (Kanaya et al. [51]). These authors performed depolarized light scattering (DPLS) measurements during the annealing of a melt of isotactic polystyrene after a pulse shear treatment. The melt was sheared at several temperatures and at a shear rate of 30 s^{-1} for some 6.7 min. At the lowest temperature of 210 °C the scattering pattern showed the occurrence of elongated particles after 10 min. These particles were oriented in the direction of the previous flow. After 30 min, however, this pattern was covered over with a pattern typical for spherulites. At higher temperatures, however, no visible patterns were obtained.

At this point it is time for a final conclusion. Apparently, nuclei grow in the direction of flow during the application of this flow. This fact leads to the establishment of a novel mechanism of flow induced nucleation in polymer melts. The assumption is that in quiescent melts, at temperatures below the melting temperature of the spherulites, there is a huge reservoir of dormant athermal nuclei. These nuclei are local alignments, which are stabilized during the process of cooling. Necessarily these nuclei have the shape of fringe micelles. Their length determines the temperature, where they become active as athermal nuclei: the longer (the more orderly) they are, the higher is the temperature, where they become active. In fact, as has been explained already in Sect. 2.3.2.3, the height of the barrier, which is built up at the beginning of lamellar growth, depends on the length (the orderliness) of the body of the micelle because of the increase of the negative internal free energy with this length. This means that the said barrier becomes always lower, if the micelles become more orderly ("longer"). By the action of flow these micelles are oriented and their tangling ends are "ironed". Their length also increases by the association of other molecules, as has been explained in the last sections.

3.3.2.4 A Conspicuous New Insight

Introducing himself into a new aspect the reader should inspect the Figs. 3.22 and 3.29 again. In both figures quite long waiting times separate the moment of the cessation of the flow from the time, when visible depolarization becomes observable. However, because of the chosen special directions of the polars one cannot learn from these experiments, whether actually anything has happened during the previous waiting times. In fact, it has meanwhile been shown that during the waiting times birefringence effects could come up, which were oriented in the previous flow direction. As a consequence these birefringence effects could not be detected with the aid of the polars in their positions parallel and perpendicular to the direction of flow. The inspiration for using interference colors in polarized white light came quite unexpected (See Janeschitz-Kriegl and Ratajski [52]). In fact, we had just purchased the Lincam machine CSS450, which was described by Mackley et al. [53].

For our purpose we modified this machine by introducing a λ-plate between polarizer, sample and analyser. With white light one gets the well-known red of first order, if no extra optical anisotropy is introduced. However, with the introduction of a small extra anisotropy in an additional position one gets a beautiful blue color, whereas in subtraction one gets a brilliant yellow color.

The Lincam machine is a parallel glass plate rotational rheometer. The rate of shear increases linearly in radial direction. At a certain radius one has (with a stationary lower plate) a certain rate of shear. With a distance of half a millimeter between the plates and a temperature of 145 °C a shear rate of 1 s^{-1} was applied for 10 s to the melt of an industrial polypropylene. A series of six photographs, depicting the temporal development of the sample after the cessation of the flow, is shown in Fig. 3.30.

At time zero after the cessation of flow one has a homogeneous red of first order. Only after six minutes a great number of little blue specks can be observed. The blue color means that the molecules, which are contained in these specks, show a preferential orientation in the previous flow direction. Interestingly, these six

Fig. 3.30 Development of the morphology for an industrial polypropylene at 145 °C after a shear treatment of 10 s at a shear rate of 1 s^{-1} (gap width: 500 μm). The waiting times are given in the pictures. The blue areas indicate fields of orientation in the previous flow direction (Fig. 2 in Ref. [54])

minutes are in the range of times, where in Fig. 3.22 the depolarization becomes noticeable. In fact, the investigated polymers are quite similar. After 17 min the blue areas have partly fused. They cover now a great deal of the whole sample. After 21 min white spots are clearly recognizable. It appeared that these spots indicated the places, where the growth of spherulites started. Finally, after 60 min, these spherulites covered the whole picture and caused complete depolarization. The cross-sections of these spherulites showed a white of higher order. Apparently, the detection of the depolarization, as shown in Fig. 3.22, is more sensitive. In fact, 1 h means 3600 s instead of only 600 s, as depicted in Fig. 3.22.

It appears that there still remain some questions to be solved. For instance, a slight slip could not be avoided in the Lincam machine at increasing shear, when a general purpose PP was used. With a less elastic material as the peroxide degraded polypropylene the machine worked quite flawless. (See Ratajski et al. [55]). With this polymer a full series of experiments could be carried out. However, the obtained photographs were less convincing. This means that Fig. 3.30 remains qualitatively of great importance even, if the corresponding series of experiments remains incomplete.

To our discomfort Fig. 3.30 seems to show that there are two different nucleation mechanisms. In fact, one first gets blue specks, and the corresponding blue areas spread rather fast, before they start to fuse. Spherulites start growing only later on a number of spots. For the first one cannot decide, whether the blue specks finally are all transformed into nuclei, on which spherulites can grow. It is imaginable that the spherulites are initiated by foreign particles, which are always present in a much lower number than local alignments. In fact, shear flow can spare those remnants of a previous network exactly along the flow lines. In this connection one is reminded of the phenomenon of row nucleation (see Keller and Machine [56]).

Anyway, the blue areas of Fig. 3.30 do not seem crystalline. They look more like gel particles, which possess an oriented interior. In this connection the reader is reminded of the end of Sect. 2.3.3, concerning Winter's gel point. In the corresponding Fig. 2.24 it is shown that in isotactic polypropylene at temperatures above 135 °C the clouds of the gel grow faster than the spherulites. This fact is in agreement with Fig. 3.30, if one admits that the interior of the gel particles is oriented as a consequence of previous shear flow.

Finally it seems worth while to discuss the differences between Figs. 3.29 and 3.30. On both figures one can observe that the occurrence of spherulites took quite a long time. But the short time results differ considerably. On Fig. 3.29 one observes that the birefringence, which is apparently caused by thin oriented threads, shows up immediately after the cessation of the flow. In contrast, one observes on Fig. 3.30 that the development of the blue specks takes more than half an hour. Apparently, we have to do with the consequences of the differences of the structures of the two polymers. In this respect one should not overlook that a high shear rate of 650 s^{-1} had to be applied to the polypropylene, which had been degraded by the application of a peroxide, whereas a shear rate of only 2 s^{-1} sufficed to generate for the general purpose PP the pictures shown on Fig. 3.30. The temperatures were almost equal.

3.3.3 Relaxation Phenomena

3.3.3.1 A Historical Review

In the course of duct flow experiments, which were carried out at Linz University with the aid of a single screw extruder, the research group at this university came across the phenomenon of the relaxation of flow induced nuclei. This insight was obtained at a surprisingly early date, i.e. before 1987 [21, 28, 35, 57]. As the pertinent results are very instructive even nowadays, they will be given here in detail. In fact, except for very recent measurements, which will be discussed below, these measurements are the only ones, which have been carried out in a critical range of temperature. This range has been characterized in the said monograph by Fig. 1.3. This is the range between the equilibrium melting point and the temperature, where the spherulites melt. In the literature the overwhelming majority of the measurements have been carried out below the temperature, where spherulites melt, because this latter temperature has erroneously been considered as the true melting point. In all experiments, which were carried out in Linz, an industrial PP (with $M_w = 290000$ and $M_n = 50000$) was used. Samples were pushed at various temperatures through ducts of varying lengths but equal rectangular cross-section of a large aspect ratio (1:10 in mm).

In one series of these experiment [57] however, the special arrangement was used, which is depicted in Fig. 3.31. The head of the extruder is at E. Between the end of the extruder screw and head E there is a zone of rest, in which the melt can

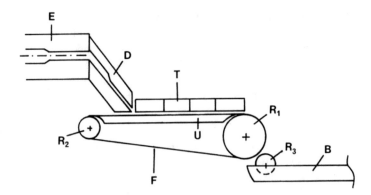

Fig. 3.31 Apparatus for the measurement of the relaxation of (lengthy) structure elements contained in the extruded strip. E extruder, D exchangeable slit die, T tunnel in segments, U heated pad, F conveyor belt, R_1, R_2, R_3 rolls of varying use, B bath (Janeschitz-Kriegl et al. [28, 57]). Courtesy of Alfred Hüthig Verlag

relax from the maltreatment in the screw channel. In part D ducts of varying lengths can be inserted. The duct exit is above the endless conveyor belt F, which is manufactured of a very thin steel band. The upper side of this conveyor belt moves in extrusion direction over two rolls R_1 and R_2. R_1 is a chill roll, which is driven by a motor. U is a heated pad and T is a heated tunnel, which consists of a number of segments. Normally the tunnel and the pad are heated to the extrusion temperature, so that the extrudate, which is deposited on the belt during extrusion, can be tempered during its movement through the tunnel, before it is quenched on the chill roll. The quench is perfected, when the extruded strip is immersed in ice water, as contained in the vessel B. R_3 is just a roll guiding the strip. In order to uncouple the residence time in the tunnel from the extrusion speed, the length of the tunnel can be varied by the use of a varying number of segments. Otherwise the residence time in the tunnel would just be inversely proportional to the extrusion speed.

If various duct lengths were used at 200 °C and the chill occurred directly after the exit of the duct, the following picture was obtained. It is reproduced in Fig. 3.32 and shows that almost always distinct surface layers are produced by the quench. The thickness of these layers depends on the extrusion speed and on the duct length. Only at very low extrusion speeds no surface layers are found. Apparently, the duct length is important. This means that the shearing time is decisive (see also the discussion of Fig. 3.19). Interestingly enough, however, the precursors, which apparently cause the oriented structure, are so tiny at 200 °C, that they cannot be traced back in a birefringence effect.

Flow birefringence, as shown in Fig. 3.33, was observed through an adapter mounted to the ends of normal ducts. This adapter contained windows in the large side walls. The lengths of the ducts were increased in this way by 7 mm. But even with the shortest duct no influence of the duct length on the flow birefringence could be observed at the extrusion temperature. This was true even for the lowest

Fig. 3.32 Thickness of highly oriented surface layers as functions of the extrusion speed (≈roller surface speed) for various duct lengths given near the curves, as obtained after quenches from 200 °C to chill roll temperatures of 100 °C (open circles), 50 °C (closed circles) 10 °C (other symbols) [21, 57]. Courtesy of Alfred Hüthig Verlag

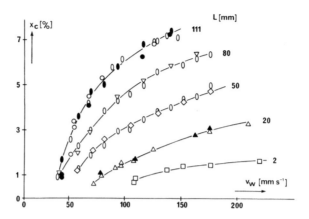

Fig. 3.33 Flow birefringence, as measured at the duct ends in the 1,3-plane (light beam normal to the large duct walls), for duct lengths quoted in Fig. 3.31. These duct lengths are increased only by 7 mm. The positions of the various symbols standing for this spectrum of lengths have no influence on the course of the curves. The extrusion temperatures are given near the curves [21, 35]. Courtesy of Steinkopff Verlag

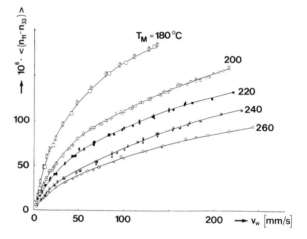

extrusion temperature of 180 °C, where the polymer melt should be much more sensitive to crystallization tendencies. By the way, the fact that the flow birefringence as such is not influenced by the duct length even with the shortest duct shows that the steady state flow profile of the unperturbed melt has already been adjusted after a very short distance from the duct entrance. This fact is in huge contrast to the presence of still invisible precursors, which become visible only after the quench. In fact, as Fig. 3.32 shows, the formation of precursors is tremendously influenced by the duct length. (In this connection, however, it should be forgotten that this figure shows overgrowth after the cessation of the flow).

By the way, from Fig. 3.32 one can also learn that the temperature of the chill role has no influence on the formation of the oriented layers. This puzzle, however, can be solved, when Fig. 3.34 is observed. This figure shows the relaxation of the precursors of PP, if the quench is postponed for several time spans of tempering in the tunnel. Except for the two uppermost curves, which were obtained, when the

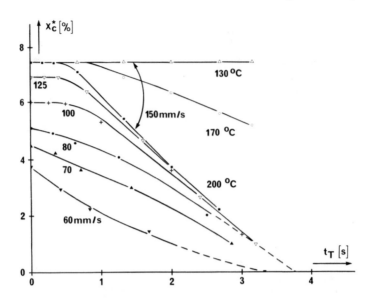

Fig. 3.34 Relaxing thickness of the surface layer as percentage of the final strip thickness, when the residence time t_T in the tunnel was varied [57]. Extrusion occurred at 200 °C through a duct of 111 mm length. Extrusion speeds and temper temperatures are given near the curves. Chill-roll temperature was 10 °C. Courtesy of Alfred Hüthig Verlag

tunnel temperature was lowered to 170 and 130 °C, all curves present results, which were obtained, when the temperature agreed with the extrusion temperature of 200 °C. For the highest extrusion speed of 150 mm/s temper temperatures of 130, 170 and 200 °C were applied. For all the other curves the temperature was kept at 200 °C, but the extrusion speeds were reduced step by step. The consequence of this stepwise reduction is clearly seen at time zero, were a stepwise reduction of the layer thickness is found. It can clearly be seen that after less than four seconds no traces of previous flow treatments can be found at 200 °C, irrespective of the previous extrusion speed. This shows that the relaxation time of the layers is not very much influenced by the prehistory. Enormous differences, however, are shown if the temper temperature is reduced. In fact, at 130 °C the layer thickness is no longer influenced by the temper time of 4 s. This fact explains, why the chill role temperatures, when varied between 90 and 10 °C, had no influence.

What can be learnt from these measurements? Apparently, flow produces a sufficient number of stretched molecules, which can associate sporadically in the temperature range between 212 °C and—say—170 °C, where metastable situations can be expected (see Fig. 1.3). Normally the range of metastable states is characterized by living equilibria, where already formed associates disintegrate spontaneously, but are replaced by other newly formed ones. Such an equilibrium can be expected without any doubt also for continued shearing. But, apparently, the steady state of this equilibrium could not be reached in the present experiments because of insufficient shearing times. On the other hand one must realize that after cessation of flow relaxation of the formed structures is inevitable. The reason for this relaxation is the fact that those molecules, which separate spontaneously from the precursors, immediately coil up in the quieted down melt and are, therefore, no longer available for the formation of new stretched nuclei. Because of the fact, however, that the stability of the formed precursors increases rapidly with decreasing temperature, this consideration loses its importance. For secondary nucleation, were only local rearrangements are required, this fact is not so obvious, as has been posed in the second chapter of this monograph (see Sect. 2.3.1).

As has been mentioned in the original papers, one cannot learn from Fig. 3.32, whether with increasing shearing time the thickness of the layers tend to a final value or not. Such a judgment is rendered difficult also by the fact that at a given extrusion speed the shear rates (and the shear stresses) decrease with the distance from the duct wall. This means that thick layers finally grow with a reduced speed. Another point is that the extruded fluid layers change their shape, as soon as they are deposited on the conveyor belt. The speed of the conveyor belt is adequately chosen: the thickness of the extruded strip must be lower than the height of the duct, from which it is protruded. This is the reason, why the thickness of the oriented surface layer is given in percentage of the obtained thickness of the strip. Fluid layers, which have previously been close to the duct wall, are considerably stretched and, as a consequence, become much thinner. In the center of the strip layers are reduced in length and become thicker. But the shortcomings are compensated by the fact that the shown results cannot be obtained otherwise. The merits of G.Krobath should not be concealed.

As reported in Ref. [21], Wippel tried to obtain some better insight into the kinetics of the formation of oriented surface layers in the said high temperature range. For the purpose he had to get rid of the uncertainties, which arose in connection with the deposition of the extruded fluid strip on the conveyor belt. The only possibility was to quench the duct as a whole and to extract the solidified content. For the purpose the walls of the duct were equipped with channels, through which cooling water could be pumped as close to the duct surface as possible. Also, the duct was slightly tapered for an easier extraction. Cross-sections were cut in the 1,2-plane. The boundaries of the highly oriented layers, as obtained for various throughputs, were plotted against the distance from the duct entrance. For one of the series of measurements these results are given in Fig. 3.35.

The flow profiles were calculated for the still homogeneous melt. In this way the residence times as functions of the distance from the entrance could be calculated for all kinds of fluid layers. These layers were located at varying distances from the duct surface and at corresponding shear rates. If at a certain distance from the duct wall a line is drawn parallel to the wall on a drawing like the one shown in Fig. 3.35, one obtains points of intersection with the boundaries of diverse oriented layers. From the distances between the entrance and these points of intersection the pertinent residence times can be obtained. At those residence times the layers were ready to form oriented structures after the quench.

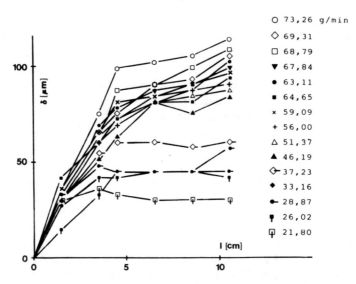

Fig. 3.35 Boundary layer thickness versus distance from the entrance to the duct, as obtained by quenches immediately after the cessation of flows at 200 °C for the PP of Fig. 3.28 [21]. The output rates are given on the right side of the figure. Courtesy of Pergamon Press

Fig. 3.36 A double logarithmic plot of the critical residence time against the corresponding shear rate for the PP of Fig. 3.28. at a temperature of 200 °C. Unpublished graph, as prepared by Dr. Herbert Wippel

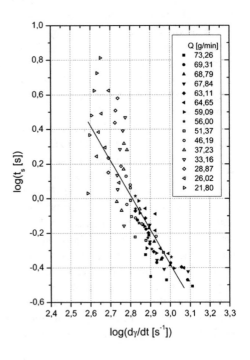

On the next graph (Fig. 3.36) a double logarithmic plot of the obtained critical residence times against the corresponding shear rates is given for all throughputs of the series of Fig. 3.35. In spite of the scatter, as caused by the complicated method, one can clearly see that the critical residence times decrease rapidly with increasing shear rates. A line of slope minus two is inserted. It must only be said that below a shear rate of 400 s^{-1} no surface layers were found. One has approximately

$$t_v \approx C q^{-2}, \tag{3.18}$$

where t_v is the critical residence time and q is the corresponding shear rate. Of course, one can rearrange this equation into $t_v q^2 = C$ and one can also take the square of this equation: $t_v^2 q^4 = C^2$. In the present experiment one cannot discern between these options. But the conclusion is that the given results are not in contradiction to the results described in Eq. (3.6) of Sect. 3.3.1.1. For the moment it is of great interest that even at the high temperature of 200 °C, where the precursors relax quite rapidly (see Fig. 3.34), their growth mechanism cannot be very different from the growth mechanism observed at much lower temperatures, where the precursors are practically stable.

A calculation of the specific mechanical work seems of interest. If one takes the point $t_v = 0.44$ s, $q = 1000$ s^{-1}, one obtains $w \approx 44$ MPa, if the viscosity for this very high shear rate is estimated from the data given in Ref. [58] ($\eta(0$ s$^{-1}) \approx 5600$ Pa.s, $\eta(1000$ s$^{-1}) \approx 100$ Pa.s). This value of 44 MPa is certainly not completely unrealistic. In fact, for the transition from the fine grained layer to the highly oriented layer, as shown in Fig. 3.19 for the same polymer at a temperature of 150 °C, a value for the specific work of about 25 MPa has been calculated. Admittedly, the slope in Fig. 3.36 does not seem too unrealistic, if the specific work is claimed to be the same for any transition to the highly oriented layer (as occurring after the quench), irrespective of the details of the previous flow condition. In fact, one has $w = \eta.q.t$, where $q^2 t = $ const for the chosen slope of two. But the viscosity η varies with the shear rate, which means that at the point $t_v = 2.5$ s and $q = 460$ s^{-1}, with $\eta \approx 166$ Pa. s, the specific work should be $w \approx 66$ MPa. This example shows that there is some uncertainty. Nevertheless, the orders of magnitude are instructive. In fact, one cannot learn very much, if—say—shear rates of the order of 1 s^{-1} or less are applied (see the majority of recently published papers).

At this point a discussion of the said "uncertainty" becomes over-due. It is conspicuous that the concept of the specific work becomes less convincing just with experiments, which are carried out at high values of the shear rate. In fact, with Liedauer's experiments [17] an interpretation of the results was possible, if the influence of the viscosity was disregarded. (See our "law", as given in Eqs. (3.6) and (3.13)). Under this condition one ends up with pure kinematics. And by this picture several facts become perspicuous: The parameter of Eq. (3.13a) appears to be almost independent of temperature and the convergence of the lines of Fig. 3.23 points in this direction. So, we can notice a situation of transition. Thus, one can also see, how difficult the theoretical description of flow induced crystallization can be in principle.

3.3.3.2 Other Ways of Observing the Relaxation of Precursors

Wippel used the decay of the birefringence of surface layers. The pertinent values of the birefringence were obtained on the quenched samples after increasing waiting times, which had been applied at the still high temperatures. In these experiments Wippel made use of a series of extrusion temperatures between 210 and 190 °C. At temperatures above 210 °C relaxation was too fast, anyway. By the way: 212 °C is the equilibrium melting point of iPP [59]. Below 190 °C problems arose with the instability of the flow at the high throughput rate of 75 g/min, which was chosen throughout this series of experiments. The shear rate at the duct wall, corresponding with this throughput, was nearly 2000 s^{-1}. In contrast to the findings of Krobath [35], which were obtained on the conveyor belt, the thickness of the boundary layer did not decrease during the relaxation of the precursors. Only the birefringence of the quenched layers decreased. Wippels results are shown in Table 3.1. Importantly enough, a single temperature dependent relaxation time was found: The curve of log Δn versus t_{rel} was found to be a straight line of negative slope within the error margin. A slight correction had to be carried out for an underground birefringence, which apparently was the consequence of the cooling process. In fact, for the samples there was a tendency to stick to the walls. A similar, more distinct experience was made later with PB-1, which clearly sticked to the wall [60].

The results of Table 3.1 were used for the preparation of Fig. 2.20 of the previous chapter because of the value of this figure in providing evidence of stable athermal nuclei at temperatures higher than the melting temperature of the spherulites. By the present Fig. 3.37 a comparison is rendered possible between the zero shear viscosity η_0 and the said relaxation time τ of the precursors. For the purpose a double logarithmic plot is prepared. If the temperature dependencies of both functions would be equal, a straight line under 45° should be expected. In reality, however, a very low slope is obtained. The temperature dependence of the relaxation time τ of the precursors is much higher, pointing to the fact that a supermolecular structure is relaxing. This fact is in contrast to the relaxation of the orientation of free or entangled molecules, which is responsible for the viscosity. In Fig. 3.37 the experimental points stand from left to right for the decreasing temperatures of the table. From an Arrhenius plot one obtains an activation energy of 334 kJ/mol. [61]. This activation energy is almost ten times the activation energy of 44 kJ/mol, which is obtained for viscous flow [63].

Table 3.1 Relaxation times of the precursors responsible for the oriented crystallization, as found at various flow temperatures for the iPP of Fig. 3.19

Temperature [°C]	Relaxation time [s]
190	57,8
195	15,88
200	7,27
205	3,67
210	0,83

Fig. 3.37 The logarithm of the zero shear viscosity versus the logarithm of the relaxation time of the precursors for oriented crystallization for the PP of Fig. 3.19 [21]. The temperatures, as applied during shearing, were from left to right 210, 205, 200, 195, 190 °C. Courtesy of Pergamon Press

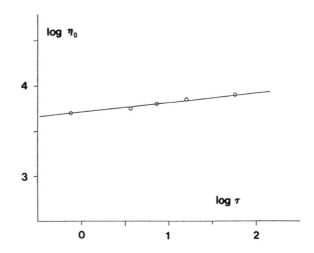

An analogical behavior was more recently found for the melt of a PB-1 [59, 60]. For this polymer it was found that under the influence of shear flow the stable crystal modification Form I immediately showed up. This fact was documented by an infrared study [60]. The difficulty with this polymer is that it attaches to the duct wall during the quench, if crystallization does not occur fast enough. Apparently this is the case, if shearing is applied at too high a temperature. As a consequence, cooling stresses and a corresponding birefringence come up because the polymer can contract only in a direction perpendicular the wall. (This is a well-known effect with the manufacture of compact discs [62]). Nevertheless, the impression was that precursors for "Form I" are formed at a temperature as high as 170 °C. This is remarkable, if one realizes that the quoted melting temperature of spherulites of Form I is between 124 °C (for the Shell copolymer used in the cited research) and 134 °C. Unfortunately, the equilibrium melting point of "Form I" seems to be unknown. The data for the investigated PB-1 are given in Table 3.2, which corresponds to Table 3.1 for PP.

For the precursors of this polymer an activation energy of 224 kJ/mol could be derived from the data of Table 3.2. For PB-1 the activation energy of viscous flow is about 44 kJ/mol, as with PP [63]. Again, one finds such an enormous discrepancy.

Table 3.2 Relaxation times of the precursors responsible for the oriented crystallization, as found at various flow temperatures for an PB-1 with the code name PB0110 of Shell [60]

Temperature [°C]	Relaxation time [s]
150	14,3
155	4,5
160	3,2
170	<0,9

On the basis of experiments, where a glass fiber was pulled through the undercooled melt, Alfonso and Azzurri of Genova   [64–66] documented the relaxation of precursors in the melt of a high molar mass PB-1. These authors observed the disappearance of the track, which such a fiber left behind. This track consisted of a transcrystalline layer, which apparently had grown on the said precursors. Apparently, these precursors were oriented in the direction of the fiber. The impressive results of these authors are shown in Fig. 3.38, which is similar to our Fig. 2.20. There are a few differences, however. Our measurements on iPP were carried out in a temperature range between 190 and 210 °C, whereas the measurements in Genova were carried out between 125 and 140 °C. Also, our figures give the relative decay whereas the holding times of Genova give the times of the total disappearance of traces. In fact, the activation energy, as calculated from our results on a similar PB-1 (see our Table 3.2) is 224 kJ/mol, whereas about 500 kJ/mol can be derived from the results shown in Fig. 3.38. Nevertheless, the conclusions can be very similar. During cooling relaxation times of the said precursors reach enormously high values, before the temperature is reached, where the spherulites of the stable modification (Form I) melt in a quiescent melt. According to the manufacturer, this temperature is 124 °C. It looks, as if the measurements, which are reported in Fig. 3.38, are less representative, because the temperature range of these measurements is more close to the melting temperature of the spherulites. However, this fact is only a consequence of the technique of the measurements: at 125 °C a holding time of 56000 s was found: This time corresponds with more than 15 h. At best one minute could be observed at 190 °C in our measurements on PP (Fig. 2.20).

Alfonso and Azzurri   [65, 66] also looked after the dependence of the said holding times on the molar masses of their PB-1 polymers. The molar mass characteristics of these polymers are given in Table 3.3.

The authors observed that the lifetimes increased with the 3.4th power of the molar mass. The fiber pulling conditions remained unchanged for this series of

Fig. 3.38 Holding time versus temperature for the disappearance of traces of a track, produced by a glass fiber, when drawn through the undercooled melt of a PB-1 [64]. Courtesy of American Chemical Society

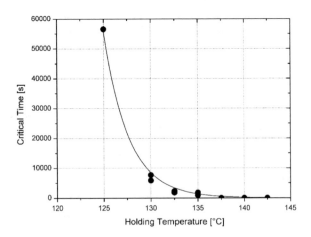

Table 3.3 Molar mass characteristics of the PB-1 samples of Shell Comp

	M_n [kD]	M_w [kD]	M_z [kD]	M_w/M_n	M_z/M_w
PB0400	31	176	368	5,7	2,1
PB0300	35	305	744	8,7	2,4
BR200	39	762	2460	19,6	3,2

measurements, namely 5 mm in 1 s at 150 °C. But this means that the only parameter, which remains of changing influence on the specific work, is the viscosity, which in fact increases with decreasing temperature with the 3.4th power of the weight average molar mass. Viewing this result from another corner of the eye one arrives at the conclusion that this result furnishes another example for the usefulness of the specific mechanical work as a decisive parameter in the field of flow induced crystallization. Having mentioned the conditions of fiber pulling one is inclined to quote an equation, which describes the shear rates near the fiber surface. This equation has been derived by Monasse [13] some time ago. It reads:

$$q(r) = \frac{1-n}{n} \frac{1}{r^{1/n}} \left[\frac{1}{r_f^{1-1/n} - R^{1-1/n}} \right] V_f, \qquad (3.19)$$

where r is the distance from the fiber axis, R is the theoretical radius of a cylindrical vessel, which stands here for half the thickness of the polymer sample, in which the fiber is embedded, r_f is the radius of the fiber and n is the power law index, describing the non-Newtonian flow behavior of the unimpaired melt and V_f is the speed of the fiber. One should probably mention that at the involved high shear rates the approach to the steady state of flow occurs in a homogeneous melt within microseconds.

Supplementary it should be mentioned that a few years after Ref. [35] had appeared, Varga and Karger-Kocsis [67] reported a relaxation phenomenon, which occurred, when a glass fiber was pulled through undercooled melts of PP and its copolymers. In this connection it must be said that quite high shear rates of the order of 100 s^{-1} can be achieved close to the fiber surface. For the homopolymer of PP a critical temperature of about 140 °C was found. Above this temperature only precursors for the α-crystal modification were formed. This fact was in accord with our earlier findings [17, 21, 35]. In fact, for this reason our experiments were carried out at temperatures between 140 and 200 °C. When in the experiments of Varga and Karger-Kocsis the temperature of the experiment was increased step by step, relaxation became faster and faster. Also this finding was in accord with our results from 1987 [57]. (see also Fig. 2.20). These facts, however, should not diminish the merits of the said authors, who reported on a host of experiences with copolymers, and on experiments at lower temperatures, where the transition to the β-crystal modification could be studied.

Very recently interesting investigations were carried out on isotactic polystyrenes [68]. The research group of Genova carried out micro-SAXS-WAXD measurements on the tracks of a fiber pulled through a melt of iPS at 260 °C. Reportedly, similar measurements were carried out by Kanaya et al. [51]. Of particular interest is in this connection that Al-Hussein and Strobl [69] have improved the determination of the equilibrium melting point of iPS. According to the linear Hoffman-Weeks extrapolation this equilibrium melting temperature should be 242 °C. But this would mean that some of the above shearing experiments were carried out above the equilibrium melting point. However, according to our experience a success of such a measurement is very improbable. So we are very glad with the finding of Al-Hussein and Strobl. These authors followed the melting and crystallization lines with the help of modern techniques in a similar way as Marand et al. [59] did for polypropylene. They arrived at the conclusion that the equilibrium melting point of isotactic polystyrene is 289 ± 5 °C. But this means that all shearing experiments were carried out also for this polymer within the temperature range given by our Fig. 1.3. (The melting temperature of the spherulites of iPS is at 230 °C).

More recently Azzurri and Alfonso reported on the relaxation behavior of shear induced nucleation precursors in samples of isotactic polystyrenes of narrow molar mass distributions [70]. In Table 3.4 the molecular characteristics of a sample of a broad molar mass distribution is given together with the characteristics of four samples of narrow molar mass distributions of increasing weight average molar masses. Because the samples of narrow molar mass distribution were very small, only the fiber pulling method could be applied to these samples. (These samples were donated by H. Marand of Virginia Tech. University). The Linkam parallel plate apparatus was applied only to the unfractionated sample.

Fiber pulling was mostly applied at a temperature $T_p = 250$ °C after a heating procedure erasing residues of previous crystallization. In a standard experiment the glass fiber was pulled with a constant speed of 5 mm/s over a distance of 5 mm. According to Eq. (3.19) this fiber speed corresponds with a shear rate of about $103\ s^{-1}$ at the surface of the fiber of a diameter of 17±1 μm. A total shear of $\gamma = 1000$ was obtained in this way in five seconds. After the arrest of the fiber the sample was transferred to one of three temperatures T_R (260, 250 and 240 °C) for relaxation at various holding times. After this procedure the sample was transferred to a temperature of 180 °C for final crystallization. This temperature was chosen because growth speed has a maximum at this temperature. Several hot stages were successively engaged for the purpose. As crystallization is very slow with iPS, there

Table 3.4 Molecular characteristics of the investigated iPS samples [70]

Sample code	$\bar{M}_n$ [kg/mol]	$\bar{M}_w$ [kg/mol]	$\bar{M}_w/\bar{M}_n$ [-]	Isotacticity index [%]
w		939	6,4	97
F1	234	291	1,24	96 ± 2
F2	563	605	1,07	96 ± 2
F3	989	1022	1,03	96 ± 2
F4	1540	1700	1,10	96 ± 2

Fig. 3.39 Fractions of surviving (point-like) nucleation precursors as functions of holding times, where several temperatures T_R were applied (after the shearing of a sample of iPS) [70]. Diamonds 260 °C, triangles 250 °C, stars 240 ° C. Courtesy of the American Chemical Society

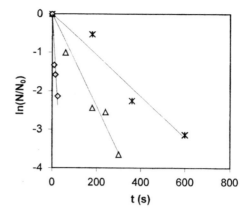

have been no quenching problems. For morphological investigation with the aid of a scanning electron microscope the sample surface was etched with amyl acetate at room temperature. At this temperature this fluid readily dissolves the amorphous phase. The etched surface was shaded with a thin layer of gold.

With the Linkam machine a gap width of 70 μm was always used. A step shear rate of $q = 30$ s^{-1} was imposed for $t_s = 10$ s. A temperature $T_s = 250$ °C was chosen for the shearing. Various holding times were applied at several temperatures, before the samples were cooled at 30 °C/min to 180 °C. Seemingly point-like nucleation centers were counted on a known area with known sample thickness. The reliability of those counts was improved by repeated counting. The number density, as obtained in permanently quiescent melts, was subtracted. In Fig. 3.39 the logarithm of N/N_0, as obtained in these experiments, is plotted against the holding time for the holding temperatures of 240, 250 and 260 °C. It can be observed that the speed of the decline increases rapidly with increasing temperature. Straight lines can be drawn through the points. This means that single relaxation times are characteristic for the pertinent relaxation processes. This is of great interest, as with this iPS the relaxation of so-called point-like nuclei is described. Previously only the relaxation of thread-like precursors (highly oriented layers) could be described for an iPP [21, 35, 57]. The fact that also so-called point-like nuclei relax according to this scheme, is important for the validity of our Fig. 1.3, which shows that with decreasing temperature one arrives at practically stable nuclei already above the temperature, where spherulites melt.

If, as in the case of PB-1, only the complete disappearance of the track of the glass fiber was considered, the authors could only find the total holding time (the "life time"), but not the course of the decay of the number density of nuclei. If the reciprocal value of this life time, which is proportional to the rate of disappearance, is plotted against the reciprocal absolute temperature ($1000/T$), one obtains an Arrhenius plot. For the samples of different molar masses these Arrhenius plots are given in Fig. 3.40. One immediately notices that the line for the polydisperse iPS (full diamonds) has the same slope and lies in the midst of the other lines. From the

Fig. 3.40 Temperature dependence of the rates of relaxation of the precursors of the investigated samples of iPS, as formed along the moving glass fiber [70]. Full triangles F4, open squares F3, full diamonds W, open circles F2, full circles F1. Courtesy of the American Chemical Society

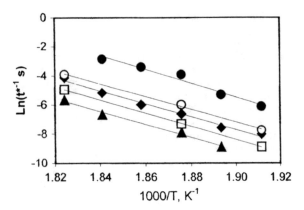

common slope one can calculate the activation energy, which is about 400 kJ/mol. This activation energy is of the same order of magnitude as with the other investigated polymers.

If on a double logarithmic plot the life times of the fractionated samples were plotted against their molar masses, a straight line was found. For a shearing temperature of 260 °C life times were found in this way, which were proportional to a power 2.2 of the molar masses. This result seems in contrast to the finding that for PB-1 samples a power 3.4 has been found. However, the commercial samples of PB-1 were polydisperse, whereas the samples of iPS were practically monodisperse. In this respect a drastic difference in the behavior of the non-Newtonian viscosity must be mentioned. With polydisperse samples the transition from the plateau at low shear rates to the power law behavior at high shear rates covers several decades in the shear rate and ends with a power law index of the order of 1/3. Also, the molar mass is still of influence in the high shear rate range. In contrast, with narrow fractions the mentioned transition from the plateau region to the power law region covers less than one decade in the shear rate and ends with a much higher power law index. Whereas the level of the plateau still obeys Flory's power 3.4 in the molar masses, the steep curves in the high shear rate range join in a common line independent of the molar masses. This fact has been shown for the first time by Stratton [71] in 1966 on monodisperse fractions of atactic polystyrenes, as obtained by ionic polymerisation. This fact means that the molar mass dependence of the viscosity of monodisperse polymers decreases rather fast with increasing shear rates. This fact may be a valid explanation for the lower power 2.2. In fact, shear induced crystallization is certainly not evoked in the low shear rate range. As the reader may notice, the present authors are defending their idea of the predominating influence of the specific mechanical work (see the explanation given above for the 3.4th power of the molar mass in the case of the PB-1 samples). In fact, if at some time in the future numerical simulations of processes will indeed become feasible, the most simple parameters will be the most successful ones. And the specific mechanical work will include also the influence of the molar masses. Exceptions will be presented mainly by samples with exotic, artificial molar mass distributions, as obtained for instance by the addition of small fractions of an extraordinarily high molar mass, as Kornfield et al. did [44] for scientific reasons. Probably, the usefulness of the specific work will also be limited at very high shear rates. A pertinent discussion has already been tried at the end of Sect. 3.3.3.1.

The recent paper by Azzurri and Alfonso [70] contains even more fascinating results. After rinsing their glass fiber with amyl acetate and shading it with gold they obtained two pictures with the aid of SEM. These pictures are reproduced here as Fig. 3.41. For the upper picture the axis of the glass fiber was in a vertical direction. The glass fiber was moved according to the recipe given above (5 mm in 5 s) at 250 °C and cooled afterwards immediately to 180 °C. On the upper picture one sees ribbed formations, which look like rolls of money. The diameter of the single coins is about 400 nm. Their thickness is about 30 nm. Probably, the molecules are oriented in tangential directions within the coins. This would mean that their average direction is in the direction of the axis of the glass fiber. The appearance of the coins reminds us of the "spokes" seen in Fig. 3.10. In this figure a cross-section perpendicular to previous flow direction is shown. These spokes may be projections of coin-like formations. The coins of Fig. 3.41, however, were formed after the arrest of the fiber as spherulites, which were spatially restricted in their lateral growth.

Fig. 3.41 a SEM micrograph of the transcrystalline layer, as formed on a glass fiber during 15 min at 180 °C in a sample of "monodisperse" iPS being quenched immediately after cessation of flow at 250 °C. The scale bar corresponds with 400 nm. **b** SEM micrograph of a macro shish-kebab, as formed during 2 h at 180 °C on a sample of iPS being quenched one minute after cessation of flow at 250 °C. The scale bar corresponds with 20 μm [70]. Courtesy of the American Chemical Society

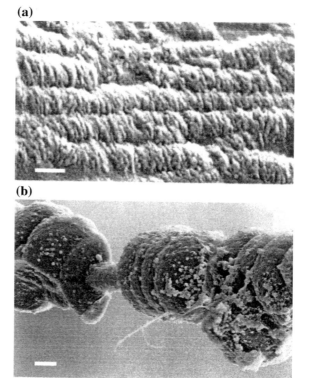

(a)

(b)

It goes without saying that the formations shown in Fig. 3.41 are the result of overgrowth on a two-dimensional periodic pattern. There is one period of about 400 nm in the previous flow direction and another period in the transverse direction of only about 30 nm. Certainly this pattern was formed during the movement of the fiber at 250 °C. The overgrowth, however, was formed only after the cessation of flow and a quench to 180 °C. If in the already quiescent melt the quench was postponed for another minute, the lower picture of Fig. 3.41 was obtained. The magnification of this lower picture is much less. Part of the fiber is seen. Here the glass fibre is in a horizontal position. One can see that the glass surface has not been active in nucleation. Apparently, the above two-dimensional pattern has largely disappeared during the rather short time of tempering at 250 °C. Probably, the character of the—practically—two-dimensional pattern has to do with the fact that with fiber pulling the shear rate decreases extremely fast with increasing distance from the fiber surface. There is only a narrow zone of sufficient shearing. The decrease of the shear rate at the wall of a duct of rectangular cross-section must be seen as very mild compared with this rapid decrease at the fiber surface.

Furthermore, the great regularity of the said pattern cannot be the result of a process governed by statistics. It seems typical for a crystallization process. But, how can it occur? Unintentionally the reader is reminded of the occurrence of the so-called long period in melt-spun synthetic fibers. This long period comes up during a second stretching of the fiber at a temperature not too far below the melting temperature. By this second stretching the modulus of the synthetic fiber is improved. One should say that the period in the direction of the glass fiber of the above experiment would point to the existence of a similar crystallization process proceeding inside the precursors along the fiber. In fact, during the initial association process of macromolecules many irregularities are unavoidably built in. In a phase of improvement these irregularities may be on the move and herded, until their concentration becomes so large that they form a separate phase. This development should start at the primary nucleus, where the thread started growing in two opposite directions. As there is no reason for variations in the orderliness of the primary thread, the critical length of the ordered sections must be uniform. This fact would explain the periodicity in the direction of the glass fiber. Apparently, the lateral growth of the precursor is coupled with the growth in the flow direction. As growth in the direction of the thread must be much faster because of the anisotropy of the associated long molecules, the lateral dimension must be much smaller. This fact seems reflected by the much smaller period of only 30 nm in the transverse direction. A lateral coupling may be the reason for the fact that all precursors are in phase with their neighbors. Such an apparent lateral coupling had already been discovered some time ago by Monasse for a high density polyethylene [72], which was sheared between parallel glass plates. Sporadically, also in an investigation at Linz University the transverse extension of a formation has been found for a polypropylene [58].

But this means that the monodispersity of the samples of iPS must not necessarily be the reason for the nice periodicity shown in Fig. 3.41. In fact, the well known row nucleation, which has frequently been found in the past, also points to a periodicity in the direction of the thread-like precursor. However, if the precursor does not exactly lie in the plain of the cross-section, one cannot see it on the picture.

Azzurri and Alfonso [70] concluded that the value of the activation energy, which was found to be the same for all samples of iPS, irrespective of their molar masses, forms a clue to the mechanism of the discussed relaxation. In this connection they quoted an interesting investigation, as carried out at Eindhoven University by Lippits et al. [73, 74]. These authors investigated the melting behavior of ultra high molecular weight polyethylenes (UHMW-PE) of molar masses above 106 Daltons. If such a polymer is polymerized with the aid of a single site catalyst at a low temperature, its nascent molecules directly crystallize during the process of polymerization. Apparently, these crystals do not contain amorphous regions, in which entanglements can be formed. The authors called this type of polymers "disentangled" nascent polymers. The crystals of those polyethylenes melt at a temperature of 141 °C, which is close to the equilibrium melting point. In a temperature range above 135 °C these crystals disintegrate rather fast. Below 135 °C, however, the rate of disintegration slows down considerably. If in Arrhenius plots the logarithms of the pertinent relaxation times are plotted against the reciprocal values of the absolute temperature, one obtains very different activation energies above and below 135 °C. Above 135 °C one finds 5000 ± 1000 kJ/ mol. Below 135 °C two activation energies of 2100 ± 150 kJ/mol and 600 ± 59 kJ/mol, respectively, were obtained. Apparently, above 135 °C the crystal lattice collapses as a whole, whereas below 135 °C two more subtle processes play a role. The lowest activation energy is interpreted as referring to the separation of single molecular segments bit for bit from the surface of the crystals. There is no doubt that the separation of a single mol of a repeating unit from the surface costs much less energy than the removal of an equal mol from the interior of the crystal. If re-crystallization is carried out after complete melting, one finds for the newly formed crystals melting temperatures near 135 °C. The above mentioned diversity of activation energies can no longer be observed. One only finds the large relaxation times corresponding with the highest activation energy.

Azzurri and Alfonso were intrigued by the relatively high activation energies of the order of 500 kJ/mol, which were found for several polymers of usual molar masses, if their precursors relaxed after the cessation of heavy flow treatments. In fact, for the precursors of the samples of iPS an activation energy of about 400 kJ/ mol was found. As one can assume that no entanglements are present in those highly oriented structures, a comparison with the behavior of disentangled nascent samples of UHMW-PE seems justified. This means that one can assume that single segments of long molecules are dispatched bit by bit during the relaxation of oriented precursors. From the point of view of the position of the equilibrium melting point, this relaxation happens always in the high temperature range of the metastable states. However, in contrast to the occurrences in the metastable regions of low molar mass compounds, where new nuclei are permanently formed in a living equilibrium, until one of them is big enough for growth, no re-association can occur in the melt of macromolecules because of the fact that those freed macromolecules directly coil up in the quieted down melt. The probability that new extended conformations are formed sporadically, is imperceptible, if there is no action of flow. Nevertheless, it is incorrect to speak of the entering into a new equilibrium, as many authors claim. In fact, even for a quiescent melt the temperature range between the high equilibrium melting point and the temperature, where spherulites melt, remains the metastable range of temperatures (see our Fig. 1.3). And the activation energy of the samples of iPS of about 400 kJ/mol shows that Al-Hussein and Strobl must be right with their high melting point of iPS of 289 ± 5 °C.

As mentioned previously in Sect. 3.2, pressurization increased the number density of nuclei considerably. In this connection also an announcement is made, which informs the reader that the authors Ma et al. [29] use pressurization instead of a quench. According to Clapeyron the melting point is increased and, in this way, the degree of supercooling is also increased. In other words, the distance between the melting point and the experimental temperature becomes larger. The authors used this fact for demonstrating that in a linear polyethylene the relaxation of thread-like nuclei, which happens after the cessation of the shear flow, can be hampered by pressurization. This polymer contained a small fraction of an ultra high molar mass polyethylene and was sheared at 135 °C.

In this paper an apparatus similar to the well-known Multi-Pass Rheometer, as described by Mackley and Hassel [75], is used. The pistons, which are located at both ends of the duct, have cross-section, which are identical to the cross-section of the duct, being a flat rectangle. So, the sample can be pressurized by an opposite movement of the pistons, but can also moved up and down in the duct by a parallel movement of the pistons. So the sample can undergo a shear treatment at a defined low pressure and can be pressurized more strongly afterwards. By the application of the pressure the melting point is increased, which means that the sample is quenched. Halfway between the pistons there are diamond windows in the duct surfaces. Through these windows wide angle X-ray patterns can be observed. In this way slender thread-like nuclei, which are formed during the flow near the duct surfaces can become noticeable by a crystallization process, which is initiated by the said quench. As this quench can also be postponed for the purpose of annealing, one can study the stability of the thread-like nuclei. The authors found that thread-like nuclei, which are too tiny for a direct observation, can survive at the temperature of previous shearing a surprisingly long period of annealing.

This observation is rather satisfying. In fact, it shows that the relaxation of thread-like nuclei also occurs in the polymer, which has the highest crystallization speed of all crystallizing polymers. One has now data of four polymers, namely HDPE, it-PP, it-PB-1 and it-PS. However, for an industrial polypropylene many corresponding results have been found already more than twenty years ago in Linz. The reader must only look at Fig. 2.20 of this edition and of the first edition of this book from 2010. It seemed almost impossible that Zhe Ma et al. should have overlooked these extended studies. It is only that Zhe Ma et al. used the WAXS pattern. In Linz the birefringence in the sample was measured after the quench. This birefringence decayed continuously with increasing waiting times. Measurements could be carried out at five different temperatures, which were between the equilibrium melting point and a temperature halfway down to the temperature, where spherulites melted. At this intermediate temperature the relaxation times of the thread-like nuclei already approached infinity. The argumentation in the paper of Zhe Ma et al. agrees almost literal with the previously given argumentation.

3.3.4 Uninterrupted Flow Treatment

3.3.4.1 Continuous Shearing

Shearing of an undercooled polymer melt up to the rapid increase of its viscosity has been the eldest method for the characterization of flow induced crystallization. In this respect the papers by the school of Maxwell [1, 76] have to be mentioned. Very recently, a painstaking study has been published in this field by Hadinata et al. [77, 78], which will serve as a basis for the present discussion. The authors investigated three samples of industrial PB-1 of differing molar masses. These are the samples PB200, PB0300 and PB0400 of Table 3.3. Remarkably, the viscosities of these samples at 150 °C are extremely different, namely 230 kPas, 11 kPas and 1.8 kPas, respectively. The melting temperature of Form II of 109.6 °C was used, when the degree of undercooling was given. The melting temperature (of the spherulites) of Form I has been reported as 121 °C.

A parallel plate rotational viscometer ("Advanced Rheometric Expansion Systems ARES") was used for the mechanical measurements and the Linkam CSS450 hot stage, as developed by Mackley et al. [53], was used for the optical measurements on the microscope. After the usual annealing treatment (for PB-1 at 180 °C), which aimed at an erasure of remains of previous crystallization, the hot samples were squeezed in the shearing units to the proper thickness. For the rheometer a thickness of 1 mm was appropriate. For the Linkam stage a thickness of only 0.07 mm had to be reached. The samples were permitted to relax, before they were cooled down to the temperatures, where the measurements should take place. At the lowest temperature of 99 °C it took about one hour, before the quiescent sample started crystallizing. For the cooling procedure only about 300 s were required.

Results, as obtained for the sample of the lowest molar mass (PB0400), are shown in Fig. 3.42 for the three temperatures of 99, 103 and 107 °C. The shear rates applied are given near the diagrams. They range from 10^{-4} s^{-1} to 3×10^{-1} s^{-1}. The normalized viscosity $\eta(t)/\eta(t = 0)$ is plotted in all graphs against the logarithm of the time t of shearing. One observes that the time of the upturn is highest for the lowest shear rates. For the temperatures of 99 and 103 °C there is almost no difference in these times for the lowest three shear rates. The conclusion by the authors has been that the averaged common time corresponds with the time of the upturn in a quiescent melt. For a more quantitative treatment times were chosen, where the reduced viscosity reached its doubled value. (See the dashed horizontal lines). One can see that with increasing shear rates the times of the upturn decrease

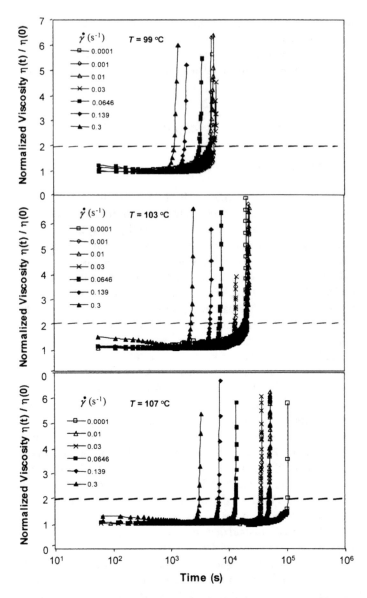

Fig. 3.42 Normalized viscosity development at crystallization temperatures of 99, 103 and 107 °C and shear rates from 10^{-4} to 3×10^{-1} s^{-1} for sample PB0400, according to [77]. Courtesy of The Society of Rheology

tremendously. By the authors these times were called the times of the onset t_{on} of the crystallization.

In Fig. 3.43 the logarithms of these onset times are plotted against the logarithms of the shear rates for the two samples of the higher molar masses and, in each case,

Fig. 3.43 The dependence of the onset time on temperature and shear rate for PB0300 (upper picture) and PB200 (lower picture), according to [68]. Courtesy of The Society of Rheology

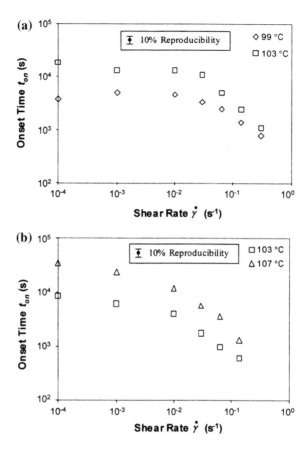

for two temperatures. The author may excuse the comment that for the sample PB200 (lower picture) the onset time for the quiescent melt does not seem to be reached at the lowest shear rate of 10^{-4}s^{-1}. Later in this section there will be an explanation for this doubt, when the morphology is considered. In fact, a look on Fig. 3.23 shows that, avowedly, for another polymer, the number density of nuclei increases at 145 °C with the third power of the specific work applied. Wolkowicz [5] has shown that also for a PB-1 the number density of nuclei increases very fast in a non-linear way with the shear rate. In that year nobody expected that the specific work was the decisive parameter. But a strong dependence on the specific work means that at equal shear rate the number density of the sample with the by far highest viscosity will be many times higher than that of the samples of much lower viscosities. In fact, at equal shear rate the specific work is proportional to the viscosity. So one can expect that the process of crystallization will be obvious much earlier in PB200. This means that one has to go for the quiescent melt to still lower shear rates for the determination of the onset time $t_{on,q}$.

Another feature of these double logarithmic graphs is that at the high end of the shear rate range the slope of the curves is minus one. This fact is particularly clear in the graph for PB0400, which is omitted here because of space economy. Beautiful curves of this type have quite recently also been published by Chen et al. [79] for an industrial polypropylene ($M_w = 3.3 \times 10^5$). These authors used also the onset time of the normal force, as measured in their cone-and-plate rheometer (Bohlin instruments), as an indicator of the onset of crystallization. Interestingly enough, in their higher shear rate range (0.07–1 s^{-1}) the onset times obtained with the aid of the normal force were a factor one half of those obtained with the aid of the shearing force (viscosity). Only at low shear rates equal values were obtained.

Hadinata et al. [78] succeeded in extending the shear rate range to values of the shear rate relevant for processing conditions. With their parallel plate machine flow became unstable at about 1 s^{-1}. In replacing this apparatus they used a concentric cylinder device and also a usual capillary rheometer. During extrusion through the undercooled capillary the effective radius of the capillary decreases with the distance from the entrance and with time. The initial extrusion pressure was calculated from the melt rheology at a higher temperature, using the time temperature shift factor. The ratio of the measured pressure and the initial pressure is shown as a function of the extrusion time in Fig. 3.44 for two apparent initial shear rates at the capillary wall, if sample PB0400 is used.

In Fig. 3.45 a double logarithmic plot is given. It shows the normalized signals, as obtained with parallel plate rheometer, concentric cylinder apparatus and capillary, as functions of the normalized shear rate. The range of slope minus one is extended here over almost four decades. The signal of the capillary occurs almost one decade later than that of the rotational viscometers. But it should be clear that always new uncrystallized melt entered the capillary. The residence times at several distances from the capillary axis take over the role of the onset times, as measured in the rotational devices. But the capillary measurements are of particular interest. They show that the highly oriented surface layers, as observed e.g. in injection molded samples, can already solidify during injection.

The reduced shear rate $q\, t_{on,q}^{0.5}\, (s^{-0.5})$ must still be explained. The authors found that by multiplying the shear rate q by the square root of the temperature dependent onset time $t_{on,q}$ of the quiescent melt, they obtained a temperature independent plot,

Fig. 3.44 Development of normalized extrusion pressure in capillary rheometer, as caused by the onset of crystallization at the wall. Apparent initial shear rates at the wall were 1 and 100 s^{-1} for PB0400 at 107 °C. Measurements were possible up to 500 s^{-1}. According to [78], courtesy of Springer Verlag

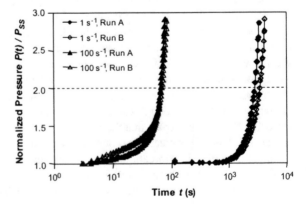

as shown in Fig. 3.45 (with the normalized onset time $t_{on}/t_{on,q}$ on the ordinate axis). The authors also mentioned in Ref. [77] that one would have got an equivalent plot, if as an abscissa the square of the said reduced shear rate would have been used, in accord with the findings of the research group in Linz [17]. In fact, the normalized onset time remains unchanged up to the moment of the upturn of the viscosity. So far this holds only for the considered special polymer, which has, as a consequence of its low molar mass, a practically shear rate independent viscosity. But the authors also tried to find a plot, which was of a more general validity independent of the molar mass of the sample. For the purpose they multiplied the above reduced shear rate by the square root of characteristic retardation time Λ. This retardation time also includes the influence of the breadth of the molar mass distribution and is defined as:

$$\Lambda = \eta_0 J_e^0, \tag{3.20}$$

where η_0 is the zero shear viscosity and J_e^0 is the equilibrium shear compliance, which does not depend very much on temperature. In this way one obtains a normalized shear rate, which reads:

$$q\, t_{on,q}^{0.5}\, \Lambda^{0.5}. \tag{3.21}$$

This normalized shear rate is dimensionless. If it is used on the abscissa, one obtains for all the three samples of PB-1 the same value of the normalized shear rate of about unity, where the deviation from the horizontal line of Fig. 3.45 starts. In Eq. (3.20) the Newtonian zero shear viscosity η_0 is used. But this means that the square of the normalized shear rate is proportional to the specific work only at low values of q (with J_e^0 as the proportionality factor). At normalized shear rates beyond unity the points for the diverse samples deviate to varying degrees from the horizontal line. The most pronounced decrease is found for the lowest molar mass. This decrease becomes less pronounced with increasing molar mass. However, proper

Fig. 3.45 Temperature invariant curves for PB0400, combining onset time data from "ARES", concentric cylinder apparatus and capillary rheometer as functions of reduced shear rate, according to [78], courtesy of Springer Verlag

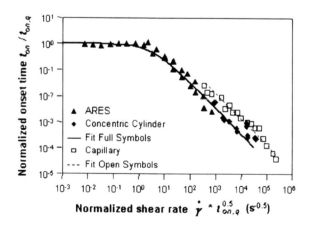

shifts to the left of the logarithms of normalized shear rates could make points of the higher molar mass samples coincide with those of the lowest molar mass. In fact the factor $q^2 t \eta_0$ is no longer a proper measure for the specific work, if the shear rate is increased. With increasing molar mass the phenomenon of the non-Newtonian viscosity becomes more and more pronounced. The viscosity decreases with increasing shear rate. Making this correction in the concept of Hadinata et al. [77] one would probably get an almost perfect reduction not only with respect to temperature, but also with respect to the molar mass.

There is still another point of great interest in the work by Hadinata et al. [77]. This interest has to do with the morphologies obtained. For two situations these morphologies are shown in Fig. 3.46 for the sample PB200 of the highest molar mass. At a temperature of 103 °C the melt of this sample was sheared for an onset time of 420 s at a shear rate of 0.139 s^{-1} (upper picture) and for a time of 100 s at a shear rate of 1 s^{-1} (lower picture). For the first mentioned treatment one finds the corresponding point in the lower graph of Fig. 3.43. It is the last point given by an open square. If a slope of minus one is drawn through this point, the corresponding straight line passes almost exactly the lower right corner of the graph. But this means that this corner should be considered as the corresponding point of the lower picture of Fig. 3.46. Rheology did not furnish this point for mechanical reasons. But the Linkam machine was able to reach the required deformation, but without enabling a viscosity measurement.

In both cases the formed morphologies could not be observed immediately after the cessation of flow. Waiting times were required for the development of these

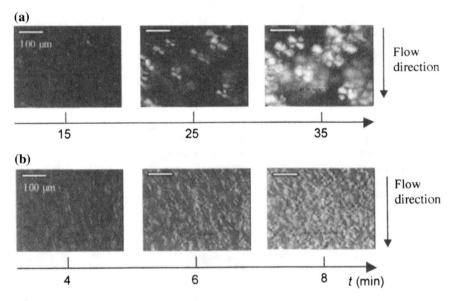

Fig. 3.46 Morphology development of PB200 after two shear treatments at 103 °C: 0.139 s^{-1} for 420 s and 1 s^{-1} for 100 s, according to [77], courtesy of The Society of Rheology

morphologies. These times are given in minutes below the photographs. Scaling bars indicating 100 μm are shown in the upper left corners of the photographs. In the waiting times the shearing times are included (7 min for the upper picture and 1.7 min for the lower picture). As also emphasized by the authors, only spherulites were found after the treatment characterized by the last point (open square—not in the corner!) in the lower graph of Fig. 3.43. But this point already lies within the range of slope minus one. Surprisingly, the upturn in the viscosity occurred, before any morphology could be observed. This means that the spherulites grow and impinge only later, after flow has been stopped. One is reminded of the unexplained mechanism prevailing in Winter's gelation (see Sect. 2.3.3). This consideration may remain also worth of a deliberation for the harsher treatment, which has led to the lower picture of Fig. 3.46. But there is another important aspect of the lower row of pictures. On these photographs one can see lengthy structure elements. But the average distance between these structure elements is of the same order of magnitude as their length. Accordingly, the transition from nuclei, which give rise to the formation of spherulites, to nuclei, which form the back-bone of thread-like precursors, occurs, when the lengths of the growing fringe micelles surpass the decreasing distances between all nuclei (between the centers of all spherulites formed afterwards). Also very satisfying is the fact that the number density between the upper and the lower row of photographs has increased tremendously.

3.3.4.2 Continuous Stretching

Also this subject has been treated in a superb way by Hadinata et al. [80]. For the purpose the "extensional viscosity fixture (EVF)" of the ARES machine was used. Between two vertically positioned cylinders rotating in opposite directions a sample of the dimensions $18 \times 10 \times 1$ mm^3 is stretched until after one revolution of the cylinders the ends of the sample meet. As the fluid sample has to be tempered and cooled, before the stretching procedure can be started at the desired temperature, sagging is a serious problem. In the EVF-unit this problem is minimized by the fact that the large surface of the sample is in the vertical position. (The width of 10 mm reduces the tendency of bending). Nevertheless, only a polymer of a sufficiently high molar mass can be investigated. Only such a polymer has a sufficiently high viscosity. As a consequence, only PB200, as characterized in Table 3.3, could be used.

There is still another experimental problem with the EVF-unit. With a parallel plate geometry, where the metal plates are in direct contact with the sample, the sample temperature can easily be controlled because of the fact that the temperature of the plates can quickly be adjusted. This is the case with shearing experiments. However, the situation with the EVF-unit is quite different. During the required cooling process the temperature of the sample can be controlled only with the aid of a stream of a gas having the desired end-temperature. This means that the adjustment of the sample temperature takes much more time. In addition, the said cylinders are heat sinks because of their axles having contact with the environment.

The authors managed to overcome all these problems. For comparison with measurements in the parallel plate unit an artificial temperature protocol was constructed for these plates. This protocol mimicked the course of the slow temperature change in the sample, which was suspended in a gas stream. In this way it could be shown that for both types of samples the upturn in the viscosity occurred at the same time, when flow was applied. This experiment showed that also in the slow cooling process premature crystallization did not occur to any extent. However, the stretching process must be carried out close enough to the melting temperature, so that crystallization is extremely slow, before flow is applied.

Additional problems can best be explained by a look on Fig. 3.47, which is taken from the paper by Hadinata et al. [80]. This figure contains two pictures. The upper one is for 101 °C, the lower one for 97 °C. In this figure the courses of the elongational viscosities are plotted against the time of stretching for various constant stretching rates. Three times the shear viscosity $\eta(t)$ was calculated from dynamic mechanical measurements at 150 °C. The obtained curve was "shifted" to the lower measurement temperatures with the aid of the activation energy of the melt. The ordinates of the obtained curves (for 101 and 97 °C) were multiplied by three. In this way one obtains the courses of the elongational viscosity, if no crystallization (and no so-called strain hardening) occurs. The courses of the shifted curves were inserted as dotted lines in both pictures. They should serve as basic lines. At 101 °C stretching rates between 10^{-4} and 10^{+1} s^{-1} were applied. A lower range from 10^{-3} to 10^{+1} s^{-1} was used at 97 °C. It is obvious that with increasing stretching rates the upturns occurred at decreasing stretching times. At 101 °C this method failed at the stretching rates of 0.01, 0.03 and 0.1 s^{-1}. The reason is that at these stretching rates the cylinders completed a whole revolution, before the upturn of the torque occurred. One notices that the experimental curves deviate from the line of $3 \times \eta(t)$ in an upwards direction, before the actual upturns occur. This deviation is ascribed by the authors to strain hardening as a consequence of entanglements. The influence of entanglements on the flow behavior in elongational flow has been treated theoretically. For the present discussion, however, it suffices, if it is mentioned that the line for the doubled value of the elongational viscosity is not drawn at proper distances above the line of $3 \times \eta(t)$, in analogy to the shear experiments, but above the sharp bents of the present experimental lines. As the onset times the times were used, where the adjusted lines for the doubled elongational viscosities cut the lines after the upturns.

For the present authors it seems of importance to point to the fact that with elongational flow all upturns occurred at times, for which the steady state of flow had not yet been reached. This is in contrast to the results on the same polymer, if shear flow is applied. With shear flow the upturns occurred much later, so that the steady state was reached before the upturns occurred [77]. Apparently the rule that steady state flow is reached already at small degrees of deformation (in shear at five units) does not hold for extensional flow.

The onset times, which have been obtained with BP200 according to the procedure described above, are shown in Fig. 3.48 for the two temperatures of 101 and 97 °C. For the higher temperature one finds a larger quasi-quiescent onset time.

Fig. 3.47 Transient elongational viscosity versus stretching time for BP200 for various strain rates at 101 °C (**a**) and 97 °C (**b**), according to [80], Courtesy of The Society of Rheology

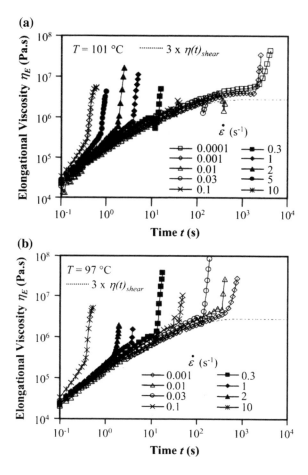

Fig. 3.48 Onset times of BP200 versus extension rates at 101 °C (upper curve) and 97 °C (lower curve), according to [80], Courtesy of The Society of Rheology

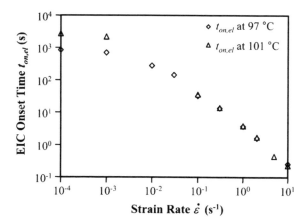

Above an extension rate of about 10^{-2} s^{-1} the influence of the temperature is no longer noticeable. In this range of strain rates the double logarithmic plot shows a slope of minus one.

If these curves are compared with the curves obtained by shearing, one observes that in the range of very low deformation rates (10^{-4} to 10^{-3} s^{-1}) corresponding points coincide. However, in the range, where the double logarithmic plots show a slope of minus one, the points from shearing lie about one decade higher. This effect becomes unmistakable, if the measurements with the coaxial cylinder apparatus [78] are included.

A temperature invariant plot similar to the one shown in Fig. 3.45 has finally been given in [80]. In contrast to Fig. 3.45 this plot, which is reproduced here as Fig. 3.49, shows the normalized onset times versus the reduced extension rates. The reduced extension rate is defined in the same way as the one for shearing. For comparison the temperature invariant results from shearing experiments are inserted. This picture clearly shows the contrast between the results in shearing and in stretching.

Fig. 3.49 Temperature invariant presentation of onset times versus deformation rates for extension experiments on PB200. For comparison the results of corresponding shear experiments are also given according to [80]. Courtesy of The Society of Rheology

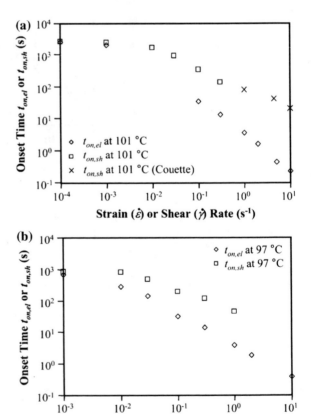

At this point the work of Okamoto et al. [81] has to be mentioned. These authors carried out an extensive study on structure development during stretching of undercooled melts of poly(ethylene naphthalates). There is one result of this work, which is of particular interest for the view of the present author. The elongational flow was realized in the elongational rheometer developed by Meissner and Hostettler [82] . Okamoto et al. succeeded to measure also the small angle light scattering, which was released by their samples. Only the polymer, which was polymerized with a Germanium catalyst, showed useful results for the present purpose. For this polymer the measurements were in the range of slope minus one in a double logarithmic plot of onset time versus extension rate. Shortly before the upturn of the viscosity a symmetric pattern was always found. Such a pattern points to the occurrence of an isotropic structure. Only after the upturn a highly asymmetric pattern was found. This pattern extended to much larger angles in the direction perpendicular to the stretch direction, indicating a larger extension of structure elements in the direction of the stretch.

There are still a few additional points, which have to be included in the discussion. It is quite clear that up to the moment, when with increasing deformation the upturn of the viscosity occurs, spherulites are formed in the arrested melt. The transformation into anisotropic structure elements only occurs after the upturn, if flow is continued. But this certainly means that solidification is not perfect at the upturn of the viscosity. This also means that the idea of a massive coil-stretch transition must be rejected. In such a case the sample should become so rigid that flow would be stopped immediately. In reality one finds that after the first coagulation of the melt spun synthetic fibers these fibers must additionally be stretched considerably in a separate process, which is known to be essential for the improvement of the mechanical strength of these fibers.

Finally, a note by Samon, Schultz and Hsiao should be mentioned [83]. In this note the authors report that the total deformation of a melt spun fiber is responsible for the solidification, irrespective of the type of polymer. In fact, chemically very different polymers were investigated, as there are: HDPE, PVDE, PA6 and POM. With all of these polymers a double logarithmic plot of crystallization onset time versus take-up speed showed a slope of practically minus one. So we can hope that an explanation of this fact can be given soon.

3.3.4.3 A Preliminary Discussion

As already mentioned, the pity of the work by Hadinata et al. has been the large degrees of supercooling. The melting temperature of its spherulites is at 135 °C. Surprisingly, the equilibrium melting point of the stable modification Form I of it-PB-1 is about 170 °C. Because of the fact that Form I appears immediately in the thread-like nuclei with shearing, one can get these 170 °C by an estimate from the relaxation behavior. This has been shown by Juliane Braun et al. [60]. (Compare Fig. 2.20, which holds for it-PP. Figure 3.38, which holds for it-PB-1, is not assertive enough because of the coarse scale of the ordinate axis, which does not permit the accurate reading of Fig. 2.20). The group in Linz has been interested in measurements, which are carried out at temperatures quite close to the equilibrium melting point. In fact, it has been shown that the two responsible processes, namely nucleation and growth, are well separated at those high temperatures. But also from a more practical point experiments at those high temperatures should be preferred. In fact, these experiments show, what happens during the cooling process, which starts at temperatures above the equilibrium melting point and is unavoidable in polymer processing. In fact, the structures, being initiated by the flow at higher temperatures, are preserved in the moldings by the rapid decrease of the temperatures.

Before a new research program can be formulated, regard has to be paid to several other experiences. The rarity of the occurrence of clicks was mentioned already, when Fig. 1.1 was scrutinized. Another point of consideration is that practically all measurements in shear flow have been carried out under steady flow conditions. In fact, according to a rule of thumb, as known to rheologists, this steady state is usually reached in polymer melts after about five shear units. An exception is made only for extremely low shear rates, which are of no interest in processing. It is quite a time ago, when the group of the present author used flow birefringence in order to verify this rule (Gortemaker et al. [84]). For the purpose the melt of a general purpose polystyrene was investigated at 170 °C.

In Fig. 3.50 the ratio of the cotangent of twice the extinction angle over the applied shear rate is plotted against the shearing time, as reckoned from the moment of the onset of the flow. For the present consideration it is only important to know that this cotangent is the ratio of the first normal stress difference over twice the shear stress. But this first normal stress difference is a measure for the free energy stored in the sample by the action of the flow. This free energy should be responsible for the crystallization tendency, as mentioned previously in this monograph. In Fig. 3.50 one finds a shaded line. This line has been calculated with the aid of Lodge's rubber-like liquid theory (Lodge [85]), using the relaxation time

Fig. 3.50 Plot of cot $2\chi/q$ versus time t for an industrial PS: Master curve according to Lodge's rubber-like liquid theory with shaded area for the limits of validity. Shear rates of the experimental curves in reciprocal seconds from above: 0.015, 0.02, 0.033, 0.05, 0.07, according to Gortemaker et al. [84], Fig. 10. Courtesy of Dietrich Steinkopff

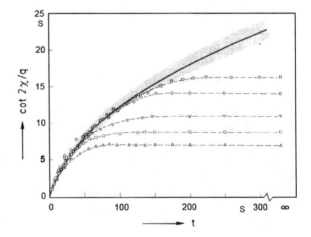

spectrum, as determined in the quiescent melt. This shaded line describes the linear case, where no dependence on the shear rate is expected for this spectrum. At the time we were very happy to be able to show that all experimental curves initially showed this linear behavior. However, all experimental curves departed earlier or later from the line for the linear case. In this respect the extremely low shear rates, where this fact could be observed, was very impressive. In fact, the applied shear rates varied between 0.015 and 0.07 s$^{-1.}$ If now the applied shear rates were multiplied with the times, where the steady state was reached, one always got a shear strain of the order five. So, the mentioned rule of thumb has actually been discovered by these experiments. Moreover, one can imagine that at the rather high shear rates, which are of interest for flow induced crystallization, the line for the ratio of the said cotangent over the shear rate practically falls into the abscissa. But this means that the entanglement network, which is responsible for the relaxation time spectrum of the quiescent melt, is completely unraveled after less than five shear units.

A supplementation to these insights has already been delivered by Fig. 3.6. For preparing this figure Liedauer had used duct flow. He measured a flow birefringence effect, which he had observed through opposite windows at the end of a long duct. In some special experiments, which were part of a larger series of experiments, he used a wall shear rate of 108 s^{-1} at a temperature of 150 °C. (See Sect. 3.3.1.1). A series of shear strains, which occurred at the duct walls, was applied. After the cessation of the flows, when the pertinent flow birefringence had disappeared, it took quite a long times, until a different type of birefringence showed up. The corresponding waiting times are plotted at the abscissa of Fig. 3.6. The shearing times, which were all extremely short compared with the required waiting times, are given in seconds at the ends of the curves. Obviously, the observed birefringence was due to the overgrowth on thread-like nuclei. But the birefringence proper, as caused by those primary thread-like nuclei, was too tiny for being observable. Figure 3.6 shows that the thread-like nuclei became observable only with the help of the overgrowth. The application of a number of shear units equal or larger than one hundred was required. These shears were obtained by multiplying 108 s^{-1} with the shearing times. But one knows that the steady state has been reached already after about five shear units. This means that the mechanism of the flow induced nucleation has nothing to do with the relaxation time spectrum of the quiescent melt.

3.3.4.4 Unlimited Shear as a Source of Information

It is obvious that pertinent experiments can be realized only with an apparatus, which permits unlimited shearing. In fact, such an experiment would widen the experimental window in the most simple way by keeping the boundary conditions unchanged. Nevertheless, the experimental realization of this concept has not been as easy. Usual rotational rheometers have appeared to be not useful. The curvature of the flow lines is too strong in those machines. As a consequence one gets secondary flow, which disturbs the required flow pattern with continued shearing. To our surprise we got this bad experience also with machines of the coaxial cylinder type. This fact became obvious already with flow birefringence measurements. But, also machines of the parallel-plate type or of the cone-and-plate type show rather large curvatures of the flow lines. There is also a second difficulty: Usually the distance between the metal walls is invariable. An exception is found only for machines, which enable the measurement of normal stress differences. In fact, axial forces are measured in those machines with the aid of the axial displacement of one of the plates. In contrast a solid distance between the walls, which is normally given, promotes slip. Actually a tendency exists with shear flow for a lateral reduction of the dimension of the sample. In this respect a big advantage is achieved with an apparatus consisting of parallel glass plates, as described earlier in this report. In this apparatus the distance between the glass plates is adaptive.

So, the decision favored a compromise. The geometry of the new machine can most easily be explained by the presentation of a picture of the sample. This sample, which is cut in the solid state from a molded plate of 1.5 mm thickness, is shown in Fig. 3.51 It has an outer diameter of 18 cm and an inner diameter of 14 cm. It fits into an apparatus of the parallel plate type. Both plates of this machine are made of aluminum. Their thickness is two millimeters. Their outer diameters are 18 cm. The lower plate is the stator. It has a ledge preventing overflow of the molten polymer. A sheet of Teflon of a diameter of 14 cm prevents the molten material from flowing into the center. The upper plate is the rotor, carrying a slight weight, which keeps the rotor in contact with the melt. The axle is centered between two journal bearings. The lower bearing is situated in a hole in the stator. The upper bearing is in the lid of the unit. There are several electric band heaters below and above the said plates.

Fig. 3.51 Ring shaped sample of iPP of an outer diameter of 18 cm, an inner diameter of 14 cm and a thickness of 1.5 mm— according to [12], Fig. 2. Courtesy of Carl Hanser Verlag

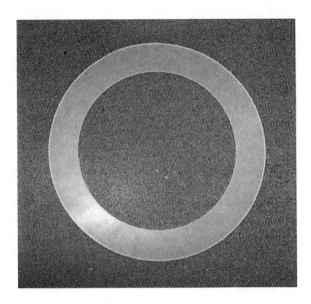

Fig. 3.52 Rotor unit with
removed front shield and
inserted copper blocks
(Fig. 5 in [12]). Courtesy of
Carl Hanser Verlag

In Fig. 3.52 a photograph is shown of the unit, when the heat shield in front is removed. Two cold copper blocks can be inserted. These blocks enable a quench of the sample. If the temperature of the shearing is close to the equilibrium melting point, this quench had to be carried out immediately after the cessation of the flow. The said blocks softly squeeze stator and rotor. This method of quenching has been advocated by Janeschitz-Kriegl et al. [86]. On Fig, 3.52 one can also see above the rotor the lower edge of a part, which carries the upper band heater. We called this part the barrel. In the picture it is tilted up in providing space for the upper copper block. During the period of shearing it is in a much lower position. During pre-heating it touches the rotor. It is tilted up only slightly, when the rotor is set in motion. A more detailed description of the unit has recently been published by Janeschitz-Kriegl, Ratajski and Eder [12].

After a period of four years of development the apparatus reached the state, in which it was ready for operation. Interestingly enough, the first experiment was already successful. Figure 3.53 shows a cross-section of the solidified sample of an industrial polypropylene. This cross-section is obtained in a plane parallel to the previous flow direction and perpendicular to the surfaces of rotor and stator. This sample was obtained after a homogeneous (!) shear strain of unheard 3200 shear units at a temperature of 195 °C, with a shear rate of 2 s^{-1} and a mild shear stress of 3.3 kPa.

For the first time in the history of flow induced crystallization a sample was obtained, where the whole cross-section was uniformly covered by a highly oriented structure. In fact, in duct flow only the zones near the duct walls have been highly oriented. It turns out that our machine is in favor of high temperatures, low shear rates and extended total shears. It was surprising that a highly oriented structure could be obtained under conditions, which had been considered as very mild from the point of view of shear induced crystallization. The clue was that a sufficiently high deformation could be achieved.

(a)   **(b)**

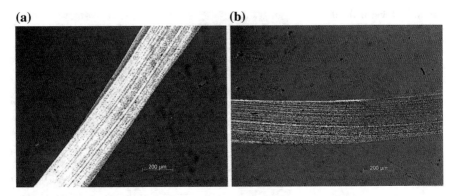

Fig. 3.53 Cross-sections through a sample of an industrial polypropylene after a shear treatment at 195 °C (see the text). The left picture is obtained under the polarizing microscope with crossed polars under 45° with the previous flow direction (**a**). The right picture shows the extinction position (**b**). (see Fig. 7 in [12]). Courtesy of Carl Hanser Verlag

Fig. 3.54 The displacement of the rotor surface is shown as a function of time at 195 °C for the used sample of iPP. (Fig. 6 in [12]). Courtesy of Carl Hanser Verlag

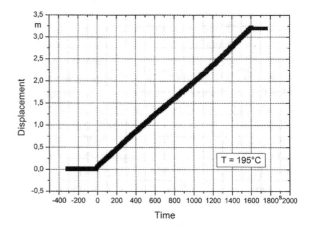

In Fig. 3.54 the course of the rotation is shown. It turned out that the flow property of the melt did not change to any extent during the period of the flow. In fact, a completely linear course could be recorded for the angle of the rotation. At the radius of 8 cm the rotor surface had undergone a displacement of 3.2 m at a sample thickness of one millimeter.

In the next figure (Fig. 3.55) a comparison is tried of the new results with those, which have been obtained some years ago on a similar polypropylene with the sliding glass plate rheometer. As was shown in Fig. 3.21, shearing times were extremely short with the latter experiments. One can read 0.15 to 0.3 s from the graph. Total shear strains between 20 and 50 were obtained, when in all experiments at a temperature of 145 °C the shear rate was 150 s^{-1} and the corresponding shear stress was 84 kPa. In the figure a double logarithmic plot is given of the number densities of nuclei, as obtained during the shearing, against the applied

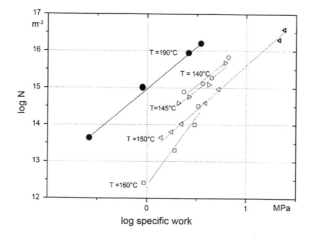

Fig. 3.55 a double logarithmic plot of the number density of nuclei versus the specific work is shown. These experiments were carried out on two industrial polypropylenes. The open symbols refer to measurements between parallel glass plates at an unchanged shear stress of 84.6 kPa, but at a series of temperatures. The closed symbols refer to measurements in the new rheometer at a shear stress of 3.3 kPa and at a single temperature of 190 °C. (Modified Fig. 9 of [12]), courtesy of Carl Hanser Verlag

specific works. These number densities were obtained by first counting the numbers of sectioned spherulites in the unit cross-sections of the solidified samples and then raising these numbers to the power 3/2. As has been observed, the number densities of nuclei, as obtained for several temperatures, tend to converge with increasing specific work. Such a tendency suggests that at 25 MPa the temperature dependence disappears. This effect would be in reasonable agreement with the conclusions, which were drawn, when Liedauer's work was discussed.

Interestingly enough, the number densities of nuclei, which were obtained with the new apparatus after the application of a low shear stress but increasingly high shear strains, are the higher ones. However, the quenches between the copper blocks reached a final temperature of about 90 °C. At this low temperature one obtains a similar number density also in the quiescent melt (see the left side of Fig. 1.1). So, in avoiding this coincidence one should discontinue the quench at a higher temperature, say at 150 °C, where the numbers of nuclei in the quiescent melt have been found to be negligible and the shear induced nuclei are already stable. Interestingly enough we have found meanwhile that a discontinuation in the cooling can easily be realized, when the copper blocks are removed at the moment, when the desired sample temperature of—say—150 °C is reached. An experimental point, which is obtained along this route, is added in the graph at a specific work of about one MPa. Actually, it seems that the shear induced nuclei already existed and started their growth, before the low temperature nuclei got a chance.

A safe conclusion can be drawn for the specific work, where for the first an oriented structure has been obtained. For the previous measurements this critical specific work was about 25 MPa. For the recent measurements it is only about 3 MPa. At the same time one can observe that the number densities of nuclei, as obtained in the new machine, are ten to hundred times as large as the number densities, which have been obtained in the old apparatus at comparable values of the specific work. Attention must be paid to the fact that the two series of measurements were carried out under very different conditions. In fact, the respective temperatures were 150 and 190 °C and the shear stresses 84 and 3.3 kPa.

One is reminded in this respect of the recent interpretation of Liedauer's results, where the product q x γ *has* been the relevant parameter. The specific work is σ x γ, where $\sigma = \eta q$, with η being the viscosity. But, if σ is kept constant, also η remains unchanged. For obtaining Liedauer's product q x γ (instead of σ x γ) one has just to divide the values of the specific works by the respective values of the viscosity.

In the present context we are interested in the ratio of the zero shear rate viscosities. With the aid of the activation energy of the melt of polypropylene of 44 kJ/mole one obtains for the zero shear viscosities the ratio $\eta(190)/\eta(150) = 0.343$. (The temperature of 150° lies between 140 and 160 °C). As the chosen specific works are comparable (see the abscissa), one must conclude that with the different shear rates applied the non-Newtonian viscosity has decreased in the old measurements much more than in the recent measurements. In other words, the non-Newtonian viscosity at 150 °C decreased much more than at 190 °C because of the higher shear rate. Because of this fact the ratio of the viscosities climbs up again to the vicinity of the value one. This fact suggests that the probability for a successful touch (a lasting "click") must have been comparable.

Most investigators still take a bet that only high shear rates cause the occurrence of oriented crystalline structures. However, there is no doubt that in the new experiments highly oriented structures are obtained already at very low shear rates. This insight is supported by the fact that during sectioning the obtained solidified samples can split spontaneously in the previous flow direction. This happened even after a mild shear treatment, when the cross-section was still covered by spherulites.

So, it is evident that one has to answer a decisive question: Is in first instance the formation of lasting clicks of importance or is the stretch of the still unbound macromolecules essential? From the measurements with the apparatus for unlimited shear we definitely learned that agglomerates of loosely assembled macromolecules were formed. They are deformed in a sea of still unbound macromolecules. During continued shear flow these still unbound macromolecules move on their practically unchanged paths. The presence of the aggregates is still of no influence on the rheology.

As soon as a macromolecule, which is brought along, is properly attached to an already existing strand, its stretch will occur within five units of strain. How-ever, the required attachment will be successful only under a precarious situation. In fact, only the attachment of one of the ends of the chain molecule to the aggregate will lead to a success. And this end must point against the local stream, which occurs at the surface of the aggregate. Any free end will drag the chain molecule loose again from the aggregate. But this means that such a successful association will be rare.

3.3.5 Adherence to the Growth Mechanism

3.3.5.1 Transformation of Dormant Nuclei by the Action of Flow [11]

In previous sections it has been emphasized that the mere orientation of macro-molecules is no sufficient condition for the formation of oriented structures. This view is confirmed, if one looks at Fig. 3.10. On this figure one observes about 15 widely separated points, where thread-like precursors have pierced the plain of the cross-section. This cross-section has been cut perpendicular to the previous flow direction. In fact, this figure caused us many years of deliberation, before we arrived at the conclusion, which will now be described. In fact, there is in principle no start of crystallization without the presence of nuclei.

But, in polymer melts there are at least two types of nuclei. One type of nuclei is presented by foreign particles, which exist as unavoidable contaminations or are introduced by a nucleation agent. The second type is born in the clean melt. For their formation flow evidently plays a role. This conclusion is underpinned by the fact that the subsequent extension of the crystalline areas, strongly depends on the flow. But it goes without saying that the sporadic formation of these nuclei is evident. One must be aware that a big part of the volume remains still free of nuclei. Because of the small initial number density of nuclei their formation occurs inde-pendently and flow is still ruled by the classical laws of polymer rheology. But these nuclei are the origins of tracks, which increase in length during the period of the flow. In the following still unbound macromolecules are associated to these tracks.. But the well-known theories for non-Newtonian flow cannot be very helpful. In these theories neighboring molecules interact only in the average and not as individuals. This kind of averaged interaction is useful above the melting point, where neighboring molecules change rapidly during flow, and there is no tendency for a permanent association.

An old question has been, how local alignments can form below the actual melting point dormant athermal nuclei in a quiescent melt. In fact, the tensions, which are caused by tangling ends of fringe micelles, depend on the lateral dimensions of these micelles. As long as the cross-section of such a micelle is very small, these tangling ends enjoy a lot of freedom (see Fig. 2.21). But this means that the effective "surface" tension at both ends of the micelle remains small. But such a positive surface tension is always in balance with a negative tension. This negative tension is a stress per unit surface in a cut, which goes through the body of the micelle perpendicular to its extension. The value of this negative tension is pro-portional to the difference of the free energies of the body of the micelle and of the fluid. The better organized a micelle is (the longer it is), the larger is the value of this difference. If the negative free energy dominates, the micelle is stable.

However, with the lateral growth of such a micelle the tangling ends get into a squeeze. In fact, the need of space in lateral direction is larger outside than inside the ordered body. The growth of the lamella will be stopped, if the barrier, as caused by the tangling ends, becomes too large. In such a case the micelle can remain a dormant nucleus. A hypothesis is now that flow irons the fringes. In fact, by the action of the flow the length of the body of the micelle can increase. But a greater length (orderliness) causes a better stability. As a consequence such a micelle can become an effective nucleus at a higher temperature. This ironing paves the way for continued longitudinal growth.

But from a macroscopic point of view it can happen that such a micelle still acts like a point-like nucleus. This will be the case as long as the length of the micelle remains smaller than the distance to the centers of neighboring nuclei. But this distance decreases because of the increasing number density of the nuclei. This means that a critical situation is reached during continued shearing. At a certain moment the length L of (a sufficient number of) micelles surpasses the mutual distance D. This fact causes in duct flow a relatively sharp boundary, which is found in the cross-sections of solidified samples between the oriented surface layer and the core. The occurrence of such a boundary is certainly furthered by the fact that the shear rate decreases in duct flow towards the central plane. The said sharp boundary can easily be localized by the fact that the lines for the changes of D and L, which occur as functions of the specific work, meet under a sharp angle (see the next section).

The just given description of the behavior of local alignments can serve also as a guide for the understanding of the following fact. On double logarithmic plots of number densities against specific work one regularly observes a slope, which is much higher than unity. (See the Figs. 3.23 and 3.24) In fact, also the early work by Wolkovicz [5] points already in this direction. It seems that in this respect the varying quality of the local alignments will be essential. In fact, these alignments will not be uniform. There will be a whole spectrum of them, and the lower their orderliness is, the higher will naturally be their frequency of occurrence. But this means that an activation of local alignments of lower quality will occur only at higher total shear strains. They need a longer phase of growth. This fact can explain the mentioned high slopes of the graphs.

For a further discussion one must in first instance realize that there are two very different time scales. First of all there is the time scale of the shearing. In the experiments, as carried out in the year 2003, shearing times varied between 0.15 and 0.3 s. After the cessation of the flow one was confronted with the rather long times needed for the growth of the spherulites in the quieted down melt. These spherulites finally covered the whole volume after their impingement. In these situations, quenches will only be necessary at high temperatures, where the nuclei start to disappear again after the cessation of the flow. In the lower temperature range of the many previous measurements, where no quenches were required, the waiting times were of the order of 100 s. But this means that all nuclei, which are activated during the very short periods of shearing (of fractions of a second), have experienced nearly the same time for their growth (of about hundred seconds). As a consequence their sizes were almost equal.

However, the transition to the occurrence of thread-like nuclei is caused by the fact that during the short period of shearing the length of these nuclei has already surpassed the distances between the centers of most of the other nuclei. After the cessation of flow the situation changes. The growth, which has occurred in the flow direction, will nearly be stopped. But lateral growth will become important. This fact caused some authors to claim that spherical cross-sections of the threads should be created. However, experience has shown that no spherical cross-sections are formed during this growth. In fact, the radius of a cylinder would increase with the square of the waiting time. If for some reason spheres were formed instead of the shishs, the pertinent radius would grow even with the third power of the time. In fact, in Sect. 3.3.1.3 it was shown that no spherical cross-sections were found on the shishs. Instead, we obtained cross-sections, which reminded us more of road-crosses on a map than of ordinary stars. They showed protruding thread-like arms. In fact, this kind of formations was in agreement with the puzzling experience that the coming up birefringence showed a linear growth with time. (see Fig. 3.10).

However, one should not forget that in both of these growth processes apparently only fractions of the nuclei were engaged. Almost misleadingly, an enormous total length of thread-like nuclei of 1.9×10^{11} m/m^3 could be calculated, if the lateral distances were used. The pertinent Eq. 3.14 has already been discussed in Sect. 3.3.1.3. But this enormous length must be divided by the usual number of nuclei per cubic meter, say of 1.9×10^{16} m^{-3}. In this way, however, one arrives at a remaining length of only micrometers. But such a theoretical length is much too short in comparison with experimentally found lengths. For instance, big experimental lengths have been obtained also by the group of prof. Kornfield [43]. The conclusion is, as already pointed out, that only a small fraction of the counted number of nuclei has caused the growth of thread-like nuclei. At first sight one could belief that only foreign particles are engaged. However, the number of foreign particles is constant and cannot show a remarkable increase, as has been observed with continued shearing.

3.3.5.2   A Continued Discussion

One of our points of consideration has been that only those nuclei can form thread-like nuclei, which have surpassed a critical length during shearing. This length must be larger than the average distance between the other nuclei. The average distance between these other nuclei is equal to the reciprocal value of the third root of their overall density. In fact, one can assume that all small sized nuclei, sometimes called point-like nuclei, grow out into spherulites after the cessation of the flow.

For the situation, as shown in Fig. 3.56, one finds in Fig. 3.57 a double logarithmic plot of the decreasing average distance D between all smaller nuclei against the growing length L of the thread-like nuclei, both being functions of the specific work applied. Line L is a border line. Above this line no lines of the same kind can be drawn. But this border line is of particular interest because of the fact that it cuts the line D under an acute angle. Actually two opposite processes occur, before these lines meet: On one hand one has the decrease of the distances between ordinary small sized nucleus, on the other hand one has the increase of the lengths of nuclei

Fig. 3.56 Cross-section in the 1,2-plane, as obtained after duct flow of a sample of iPP (Mw = 322.000, Mn = 47.000) at a temperature of 150 °C with a shearing time of 2.45 s, a shear rate at the duct walls of 169 s^{-1} and a shear stress at the walls of 0.078 MPa. Crossed polars under 45° (Liedauer et al., [38]). Courtesy of Hanser Verlag

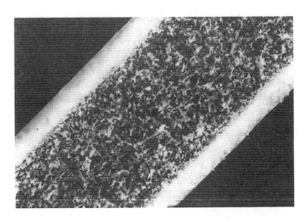

Fig. 3.57 A double logarithmic plot of D and L (in μm) against the specific work σ(kPa), as imposed at 150 °C on the melt of an industrial polypropylene at a shear stress of 84.6 kPa. This shear stress held for the wall of the duct, before the flow was stopped (see Fig. 3.23). Courtesy Hanser Verlag

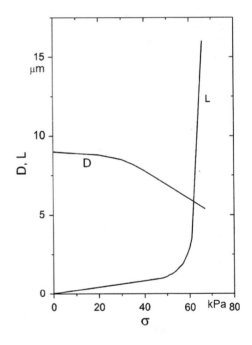

predestined to form thread-like nuclei. The intersection of these lines at an acute angle means that the transition of the pertinent structures does occur sharply.

In fact, after duct flow it was found regularly that there was a sharp boundary between the oriented zone near the duct wall and the spherulitic interior of the sample. It surprised the present authors for many years that nobody had ever mentioned explicitly of this conspicuous fact.

But here we have an explanation.

In this connection one should also refer to our observations, which were made, when Fig. 3.19 was presented. Actually, this figure is only one of three figures.

After a somewhat lower shear stress at the wall only a fine grained layer was obtained. After a somewhat higher shear stress at the wall only a highly oriented layer with a sharp boundary was found. This later figure is reproduced here as Fig. 3.56. Here one must point to the fact that the pictures shown in the last two figures are obtained only, if the cooling to the temperature of 150 °C has occurred first and the flow in the duct of rectangular cross-section has occurred afterwards at the lowered temperature. With realistic processing, however, shearing always happens first, i.e. during the cooling from the high temperature above the equilibrium melting to the temperature, where crystallization can occur—usually during increasing pressure. Obviously, this is the first time that the pressure is increased at a temperature, which is low enough for the onset of crystallization. But under those conditions separate oriented layers at the duct walls have never been found. So it must be admitted that the said last two figures are only by-products. They were obtained, when the crystallization had to be investigated at an invariable intermediate temperature in avoiding the precarious conditions prevailing during a realistic continuous cooling process. However, one must concede that the corresponding shear rates differed enormously in the said three experiments.

There are also other interesting problems. In this connection the reader is reminded of Sect. 3.3.2.4, where also two different types of nuclei are found. This finding was the result of the application of interference optics. A λ-plate was introduced into white polarized light. One has here a typical example for the difficulties in understanding, as caused by the limitations of the ranges for experimental investigations. In fact, there is no apparatus, which can measure all phenomena. The Linkam apparatus could be modified as described. But the configuration of this apparatus is not useful for the experiments, which have been carried out with our recent apparatus for prolonged shearing. But this latter machine cannot easily be modified for the introduction of interference optics. With this latter machine we were glad to be able to meet all requirements, which had to do with heat transfer and rheology. In fact, we knew in advance about the advantage of the optical approach, but we did not see a chance to combine the advantages of both approaches. A very complex machine would have been the consequence.

But this example should be a warning to those, who immediately start with mathematics and are willing to carry out only some supplementary experimental measurements or to try the interpretation of already available results, which usually have been obtained in a too narrow experimental window. One should start in the inverse direction: To invent first some new and uncommon experiments and then try to interpret the obtained results with the aid of some theory.

But there are still more problems. In fact, when the growth of oriented structures was described in Sect. 3.3.1.2, the assumption was made that there existed a linear dependence on the shearing time. One arrived at the surprising parameter $q^4 x\, t^2$. In fact, this parameter is the square of $q^2 x\, t$. The occurrence of the squared parameter is verified in Fig. 3.8. It was caused by the superposition of two processes.

However, quite recently a new insight was obtained. In fact, one can rewrite the parameter $q^2\, x\, t$ as $q\, x\, q\, x\, t = q\, x\, \gamma$, where γ is the ultimate shear strain and the remaining q is a measure for the frequency of successful touches, as occurring during the unit of the shear strain. The question remains about the birth of the starting points. In fact, the introduction of the third power of the shear strain was dictated by the experiments, which showed the occurrence of small sized nuclei (see Fig. 3.24), which would grow out into spherulites later. However, should this third power also lead to an acceptable result for the starting points of thread-like nuclei?

In order to investigate this question our Eq. 3.9 is interpreted again by separating one q and combining this q with t in order to obtain $q \times t = \gamma$. In this way the first power in γ is retained. However, instead of the first power of this factor γ its third power γ^3 is now introduced for trial. By this modification the next higher approximation, which is permitted, is introduced (i.e. $\gamma^{3)}$. In fact, the introduction of the second power of γ is not permitted. This introduction would violate the condition, that the results must be independent of the direction of the shearing. Remarkably, such a quite general consideration supports the experimental experience. As γ is proportional to t, the integral of Eq. 3.10 must only slightly be modified. If the required integration over γ^3 is carried out and the result is transformed back to the use of t, one obtains instead of $q^4 t^2$ (see Eq. 3.6) another combination of parameters, namely:

$$L \propto q^6 x \, t^4. \tag{3.22}$$

However, this combination is not useful. In fact, it would cause in Fig. 3.8 a much too steep negative slope in $\log t$, namely -4 instead of -2. This negative interpretation becomes more lucid, if one realizes that the optical retardations, as plotted on the ordinate axis of Fig. 3.8, are proportional to the reciprocal values of the lengths L of the thread-like nuclei. So it seems as if two interpretations are possible. However, quite recently the author has discovered that there is an interrelation between these ways of interpretation.

In fact, one of these interpretations holds for the quiescent and the slowly moving melt. The other one is obviously valid only under the influence of shearing. In fact, the occurrence of the first power seems more acceptable from the point of a naive view. The product of the first power of the shear rate with the obtained shear is an even function of the shear rate. (In fact, the shear itself is a product of the shear rate and the time of shearing). One gets an even function as well, if the third power of the shear rate is combined with the shear. An undesirable situation, however, is created, if the second power of the shear rate is combined with the shear. In such a case one gets an odd function of the shear rate. Such a combination would mean that the result should depend on the direction of the shearing. However, such a situation is not acceptable. As a consequence only the third power of the shear rate provides the next step of approximation. Responsible for this uncommon behavior is apparently, that the first power is blocked for some reason. Apparently only a relatively strong shear flow releases this blockade. However, as long as this blockade works, one obtains with increasing shear rate only a dramatic increase of the number of nuclei for the later growth of spherulites.

In this way it is shown that there must be a molecular mechanism behind these results. Unfortunately, however, our kinetic results cannot reveal such a process.

One would need the inspiration of a theorist. But one thing is clear: both processes must start at adequate nuclei, which are in the fluid at great distances from each other in the beginning. Nevertheless, there are some helpful considerations. For instance: the state of the stresses is linked to the state of the orientation by the factor kT (see the principles of rubber elasticity). In fact, our melts are "rubber-like liquids". But this factor causes only a mild depends on temperature, as the relative changes of the absolute temperature are small for the range of rheological experiments.

The probability for encounters of macromolecules, which lead to a "click" (see the mechanisms of crystallization), seems of central importance. In this respect the work of Prof. Tanner's group [87] in Sydney provides much hope. These authors were the first and only investigators, who attacked the problem of a real interaction between neighboring molecules. These authors treated the quiescent and the moving melt. Molecular dynamics were applied. So far, however, the ranges of validity of the results of Sydney and of our experiments are still too far from each other.

Admittedly, there will still be a formidable task. A look on the Tables 1.1 and 1.2 will not promote optimism. In fact, these tables show the tremendous influence of diverse back-bone structures on the crystallization speeds of a number of current polymers in their quiescent melts. If one selects three polymers with a C-C backbone, say, polyethylene, isotactic polypropylene and isotactic polystyrene, one finds a factor in their speeds of crystallization of six decades, with polypropylene in the middle. It is also perplexing that isotactic polystyrene, which forms the most stable crystals, is at the same time the slowest with respect to the process of the crystallization. This fact means that one has to do with a dominant influence of the kinetics. Apparently, the bulky phenyl side group cannot move easily.

The unavoidable existence of many particle problems is exaggerating the difficulties. So it may be that one has just to continue with the use of adjustable parameters. But for this purpose a series of measurements, like those carried out at the time by Liedauer, will be required. Unfortunately this means that one can become pessimistic about the benefit of our research. The answer can only be that knowledge about the actual situations will be useful anyway. It will certainly be better than ignorance, which unavoidably must lead to pure trial and error.

References

1. Haas TW, Maxwell B (1969) Effects of shear stress on the crystallization of linear polyethylene and polybutene-1. Polym Eng Sci 9:225–241
2. Mackley MR, Keller A (1973) Flow induced crystallization of polyethylene melts. Polymer 14:16–20
3. Sherwood CH, Price FP, Stein RS (1978) Effect of shear on the crystallization kinetics of poly (ethylene oxide) and poly(ε-caprolactone) melts. J Polym Sci Polym Symp 63:77–94
4. Ulrich RD, Price FP (1976) Morphology development during shearing of poly(ethylene oxide) melts. J Appl Polym Sci 20:1077–1093

5. Wolkowicz MD (1978) Nucleation and crystal growth in sheared poly(1-butene) melts. J Polym Sci Polym Symp 63:365–382
6. Van der Vegt AK, Smit PPA (1967) Crystallization phenomena in flowing polymers. Soc Chem Ind London Mongr 26:313–326
7. Devaux N, Monasse B, Haudin JM, Moldenaers P, Vermant J (2004) Rheooptical study of the early stages of flow enhanced crystallization in isotactic polypropylene. Rheol Acta 43:210–222
8. Pogodina NV, Lavrenko VP, Srinivas S, Winter HH (2001) Rheology and structure of isotactic polypropylene near the gel point: quiescent and shear induced crystallization. Polymer 42:9031–9043
9. Wereta A, Gogos CG (1971) Crystallization studies on deformed polybutene-1 melts. Polym Eng Sci 11:19–27
10. Janeschitz-Kriegl H, Ratajski E, Stadlbauer M (2003) Flow as an effective promotor of nucleation in polymer melts: a quantitative evaluation. Rheol Acta 42:355–364
11. Janeschitz-Kriegl H, Ratajski E (2005) Kinetics of polymer crystallization under processing conditions: transformation of dormant nuclei by the action of flow. Polymer 46:3856–3870
12. Janeschitz-Kriegl H, Ratajski E, Eder G (2014) Unlimited shear as a source of information in polymer melt processing. Int Polym Proc 29:402–411
13. Monasse B (1992) Polypropylene nucleation on a glass fiber after melt shearing. Mater Sci 27:6047–6052
14. Kantz MR, Newman HD, Stigale FH (1972) The skin-core morphology and structure properties relationship in injection molded polypropylene. J Appl Polym Sci 16:1249–1260
15. Mencik Z, Fitchmun DR (1973) Texture in injection molded polypropylene. J Polym Sci Polym Phys Ed 11:973–989
16. Tadmor Z (1974) Molecular orientation in injection molding. J Appl Polym Sci 18:1753–1772
17. Liedauer S, Eder G, Janeschitz-Kriegl H, Jerschow P, Geymayer W, Ingolic E (1993) On the kinetics of shear induced crystallization in polypropylene. Int Polym Proc 8:236–244
18. Kumaraswamy G, Verma RK, Kornfield JA (1999) A novel flow apparatus for investigating shear-enhanced crystallization and structure development in semicrystalline polymers. Rev Sci Instr 70:2097–2104
19. Flory PJ (1947) Thermodynamics of crystallization in high polymers. J Chem Phys 15:397–408
20. Gaylord RJ, Lohse DJ (1976) Morphological changes during oriented polymer crystallization. Polym Eng Sci 16:163–167
21. Eder G, Janeschitz-Kriegl H, Liedauer S (1990) Crystallization processes in quiescent and moving polymer melts under heat transfer conditions. Progr Polym Sci 15(629–714):678
22. Janeschitz-Kriegl H (1983) Polymer melt rheology and flow birefringence. Springer, Berlin, pp 46,63,113,146,175
23. Wales JLS, Philippoff W (1973) Anisotropy of simple shearing flow. Rheol Acta 12:25–34
24. Bandrup J, Immergut EH (eds) (1975) Polymer handbook 2nd edn. Wiley, New York, p. V-16
25. Münstedt H, Laun HM (1979) Elongational behavior of a low density polyethylene melt. II. Transient behavior in constant stretching rate and tensile creep experiments. Comparison with shear data. Temperature dependence of the elongational properties. Rheol Acta 18:492–504
26. Wales JLS (1976) The application of flow birefringence to rheological studies of polymer melts. Doctoral thesis, Delft University Press
27. Brochard-Wyart F, de Gennes PG (1988) Ségrègration par traction dans un homopolymere. CR Acad Sci Paris II 306:699–702
28. Eder G, Janeschitz-Kriegl H (1997) Processing of polymers 5: crystallization. Mat Sci Techn 18:269–342
29. Ma Zha, Balzano L, Peters GWM (2012) Pressure quench of flow-induced crystallization procurcors. Macromolecules 45:4216–4224
30. Boon J, Challa G, Van Krevelen DW (1968) Crystallization kinetics of isotactic polystyrene II: Influence of thermal history on number of nuclei. J Polym Sci A-2 6:1835–1851

31. Van Krevelen DW (1978) Crystallinity of polymers and the means to influence the crystallization process. Chimia 32:279–294
32. Van Krevelen DW (1990) Properties of polymers, 3rd edn. Elsevier, p 592
33. Keller A, Kolnaar HWH (1997) Processing of polymers 4: Flow-induced orientation and structure formation. Mat Sci Techn 18:189–268
34. Mandelkern L (2004) Crystallization of polymers, vol. 2, 2nd edn. Cambridge University Press, p. 372
35. Eder G, Janeschitz-Kriegl H, Krobath G (1989) Shear induced crystallization, a relaxation phenomenon in polymer melts. Progr Colloid Polym Sci 80:1–7
36. De Gennes PG (1982) Kinetics of diffusion controlled processes in dense polymer systems. II Effect of entanglements. J Chem Phys 76:3322–3326
37. Doi M, Edwards SF (1986) The theory of polymer dynamics. Claredon Press, Oxford
38. Liedauer S, Eder G, Janeschitz-Kriegl H (1995) On the limitations of shear induced crystallization in polypropylene melts. Int Polym Proc 10:243–250
39. Kimata S, Sakurai T, Nozue Y, Kasahava T, Yamaguchi N, Karino T, Shibayama M, Kornfield JA (2007) Molecular basis of the shish-Kebab morphology in polymer crystallization. Science 316:1014–1017
40. Kumaraswamy G, Verma RK, Issian AM, Wang P, Kornfield JA, Yeh F, Hsiao BS, Olley RH (2000) Shear-enhanced crystallization in isotactic polypropylene part 2. Analysis of the formation of the oriented "skin". Polymer 41:8931–8940
41. Kumaraswamy G, Issian AM, Kornfield JA (1999) Shear enhanced crystallization in isotactic polypropylene. 1. Correspondence between in situ rheo-optics and ex situ structure determination. Macromolecules 32:7537–7547
42. Williams ML, Landel RF, Ferry JD (1955) Temperature dependence of relaxation mechanisms in amorphous polymers and other glass forming liquids. J Am Chem Soc 77:3701–3707
43. Kumaraswamy G, Kornfield JA, Yeh F, Hsiao BS (2002) Shear-enhanced crystallization in isotactic polypropylene. 3. Evidence of a kinetic pathway to nucleation. Macromolecules 35:1762–1769
44. Seki M, Thurman DW, Oberhauser JP, Kornfield JA (2002) Shear-mediated crystallization of isotactic polypropylene: the role of long-chain chain overlap. Macromolecules 35:2583–2594
45. De Gennes PG (1979) Scaling Concepts in Polymer Physics. Cornell University Press
46. Jeffrey GB (1922) The motion of ellipsoidal particles immersed in a viscous fluid. Proc Roy Soc London 102:161–179
47. Bird RB, Armstrong RC, Hassager O (1987) Dynamics of polymeric liquids, vol 1, 2nd edn. Wiley, New York, p 171
48. Stadlbauer M, Janeschitz-Kriegl H, Lipp M, Eder G, Forstner R (2004) Extensional rheometer for creep flow at high tensile stress. part I description and validation. J Rheol 48:611–629
49. Stadlbauer M, Janeschitz-Kriegl H, Eder G, Ratajski E (2004) New extensional rheometer for creep flow at high tensile stress. Part II. Flow induced nucleation for the crystallization of iPP. J Rheol 48:631–639
50. Eder G, Janeschitz-Kriegl H, Ratajski E (2006) Towards the prediction of structure development in injection molded semicrystalline polymers. In: Greener J, Wimberger-Friedl R(eds) Precision injection molding. pp 137–152. Carl Hanser Verlag Munich
51. Kanaya T, Takayama Y, Ogino Y, Matsuba G, Nishida K (2004) Process in understanding of polymer crystallization. In: Reiter G, Strobl G (eds)Springer, Berlin, pp 87–96
52. Janeschitz-Kriegl H, Ratajski E (2014) Flow-induced crystallization in polymer melts: how Winter's gelation concept fits into the picture. Polym Bull 71:1197–1203
53. Mackley MR, Wannaborworn S, Gao P, Zhan F (1999) The optical microscopy of sheared liquids using a newly developed optical stage. J Microsc Anal 69:25–27
54. Janeschitz-Kriegl H, Ratajski E (2011) Crystallization in polymer melts: metamorphism of flow induced nuclei. Int Polym Proc 26:460–463
55. Ratajski E, Janeschitz-Kriegl H (2014) How to determine high growth speeds in polymer melt processing. Colloid Polym Sci 274:938–951

56. Keller A, Machin MJ (1967) Oriented crystallization in polymers. J Macromol Sci B1:41–91
57. Janeschitz-Kriegl H, Wimberger-Friedl R, Krobath G, Liedauer S (1987) On the formation of layer structures in plastic parts (in German). Kautschuk + Gummi. Kunststoffe 40:301–307
58. Jerschow P, Janeschitz-Kriegl H (1997) The role of long molecules and nucleation agents in shear induced crystallization of isotactic polypropylenes. Int Polym Proc 12:72–77
59. Marand H, Xu J, Srinivas S (1998) Determination of the equilibrium melting temperature of polymer crystals: Linear and non-linear Hoffman-Weeks extrapolation. Macromolecules 31:8219–8229
60. Braun J, Wippel H, Eder G, Janeschitz-Kriegl H (2003) Industrial solidification processes in polybutene-1. Part II-Influence of shear flow. Polym Eng Sci 43:188–203
61. Janeschitz-Kriegl H, Eder G (2007) Shear induced crystallization, a relaxation phenomenon in polymer melts: a recollection. J Macromol Sci Part B 46:1–11
62. Wimberger-Friedl R (1996) Molecular orientation in polycarbonate induced by cooling stress. Int Polym Proc 11:373–382
63. Van Krevelen DW (1990) Properties of polymers, 3rd edn. Elsevier, p 469
64. Alfonso GC (1999) Formation of cylindritic morphology in melt-sheared it-polybutene-1. Polym Mat Sci Eng 81:330–331
65. Alfonso GC, Azzurri F (2001) Shear enhanced polymer crystal nucleation: Interaction between molecular characteristics and flow. In: Conference flow induced crystallization of polymers, Salerno
66. Azzurri F, Alfonso GC (2005) Lifetime of shear-induced crystal nucleation precursors. Macromolecules 38:1723–1728
67. Varga J, Karger-Kocsis J (1996) Rules of supermolecular structure formation in sheared isotactic polypropylene melts. J Polym Sci Part B Polym Phys 34:657–670
68. Garcia Gutierrez MC, Alfonso GC, Rickel C, Azzurri F (2004) Spatially resolved flow-induced crystallization precursors in isotactic polystyrene by simultaneous small- and wide-angle X-ray microdifraction. Macromolecules 37:478–485
69. Al-Hussein M, Strobl G (2002) The melting line, the crystallization line and the equilibrium melting temperature of isotactic polystyrene. Macromolecules 35:1672–1676
70. Azzurri F, Alfonso GC (2008) Insights on formation and relaxation of shear-induced nucleation precursors in isotactic polystyrene. Macromolecules 41:1377–1383
71. Stratton RA (1966) The dependence of non-Newtonian viscosity on molecular weight for "monodisperse" polystyrenes. J Colloid Interface Sci 22:517–530
72. Monasse B (1995) Nucleation and anisotropic crystalline growth of polyethylene under shear. J Mat Sci 30:5002–5012
73. Lippits DR, Rastogi S, Höhne GWH (2006) Melting kinetics of polymers. Phys Rev Lett 96:218303-1–218303-4
74. Lippits DR, Rastogi S, Höhne GWH, Mezari B, Magusin PCMM (2007) Heterogeneous distribution of entanglements in the polymer melt and its influence on crystallization. Macromolecules 40:1004–1010
75. Mackley MR, Hassell DG (2011) The multipass rheometer, a review. J Non-Newton Fluid Mech 166:421–456
76. Lagasse RR, Maxwell B (1976) An experimental study of the kinetics of polymer crystallization during shear flow. Polym Eng Sci 16:189–199
77. Hadinata C, Gabriel C, Ruellmann M, Laun HM (2005) Comparison of shear-induced crystallization behavior of PB-1 samples with different molecular weight distribution. J Rheol 49:327–349
78. Hadinata C, Gabriel C, Ruellmann M, Kao N, Laun HM (2006) Shear-induced crystallization of PB-1 up to processing relevant shear rates. Rheol Acta 45:539–546
79. Chen Q, Fan Y, Zheng Q (2006) Rheological scaling and modeling of shear-enhanced crystallization rate of polypropylene. Rheol Acta 46:305–316
80. Hadinata C, Boos D, Gabriel C, Wassner E, Rüllmann M, Laun HM (2007) Elongation-induced crystallization of high molecular weight isotactic polybutene-1 melt compared to shear-induced crystallization. J Rheol 51:195–215

81. Okamoto M, Kubo H, Kotaka T (1998) Elongational flow-induced crystallization and structure development in supercooled poly(ethylene naphthalate). Macromolecules 31:4223–4231

82. Meissner J, Hostettler J (1994) A new elongational rheometer for polymer melts and other highly viscoelastic liquids. Rheol Acta 33:1–21

83. Samon JM, Schultz JM, Hsiao BS (2002) Structure development in the early stages of crystallization during melt spinning. Polymer 43:1873–1875

84. Gortemaker FH, Hansen MG, de Cindio B, Laun HM, Janeschitz-Kriegl H (1976) Flow birefringence of polymer melts: application to the investigation of time dependent rheological properties. Rheol Acta 15:256–267

85. Lodge AS (1964) Elastic liquids. Academic Press, New York

86. Janeschitz-Kriegl M, Janeschitz-Kriegl H, Eder G, Forstner R (2006) Heat transfer through metal walls of finite thickness. Int Polym Proc 21:41–48

87. Zheng R, Tanner RI, Xi-Jun Fan (2011) Injection molding. Springer, Berlin, p 64 Also personal communication

Chapter 4
Closing Remarks

4.1 Quiescent Melts

The field of polymer crystallization is already an old field and, as such, a rather extended one. As a consequence, one unfortunately cannot feel competent in all aspects. In fact, one can be impressed by the amount of special work, which has been presented so far. This certainly is the reason, why one can still feel like a newcomer, even if one has entered the field already 30 years ago for an investigation of structure formation during processing of semi-crystalline polymers. Previously, the author has been engaged by virtue of life in several other areas of polymer science, as there were: regenerated cellulose, flow birefringence of polymers in solution (development of a flawless apparatus as a starting point), single screw extrusion and polymer melt rheology (heat transfer and flow, inauguration of the flow birefringence of polymer melts). This information must serve as an excuse, some contributions in the field of polymer crystallization may have been overlooked. But this is the risk if one starts on a new subject at an advanced age. In such a case only one option remains, namely to promote one's own ideas without much delay. Of course, an incontestable condition is the conviction that these ideas are essential, original and not yet promoted by others.

One example is the allusion to the strange temperature dependence of the crystallization kinetics of quiescent melts. Nobody has ever expressed openly his surprise about the enormous difference between the equilibrium melting point and the temperature, where the spherulites of such a polymer are melting. Recently, Chen et al. [1] have friendly spoken of the "demarcation", as discovered by the present author. This demarcation is illustrated in Fig. 1.3 for iPP. The difference between the equilibrium melting point of the α-crystal-modification and the temperature range, where the spherulites of this modification melt, is more than 40 °C. A similar difference has recently been found for iPS by Al-Hussein and Strobl [2]. Also for Form I of iPB-1 such a considerable difference can be expected, as will be shown in the next section. In this connection it must be admitted that the

© Springer International Publishing AG 2018
H. Janeschitz-Kriegl, *Crystallization Modalities in Polymer Melt Processing*,
https://doi.org/10.1007/978-3-319-77317-9_4

equilibrium melting point cannot easily be obtained in quiescent melts. A complicated extrapolation method is required for the purpose, as Marand and coworkers have shown. In fact, the molecules should be associated over a sufficient length in the created crystals [3]. In reality, however, much shorter lengths are obtained, when realistic cooling speeds are applied. But the consequence of this fact is that much lower melting points are normally obtained in quiescent melts. The apparent melting points depend actually to some extent on the cooling speed, but not on the molar mass [4, 5] of the sample. In this respect one can concoct that after a certain length of association the required internal rotations around chemical bonds in the participating macromolecules have created enough internal stresses in their contiguous sections for preventing a further association of repeating units.

But this means that the usual equations for sporadic primary nucleation, as formulated for the first time by Becker and Döring [6], will not be applicable. A look on Fig. 1.3 will help us a little further. In this figure a hatched area is shown. This area indicates the presence of stable nuclei already above the temperature range, where spherulites melt. These nuclei have their origin in self-nucleation or in a flow induced process. The pertinent events have been described in previous sections. The stability of these nuclei and of other ones, which are present at temperatures below the melting temperature of the spherulites, has been explained in these sections. For these nuclei, which have the shape of fringe micelles, their thickness determines the value of the dispersing tension at the ends of their bodies. In fact, in very thin micelles the tangling ends possess a large degree of conformational freedom, which is reduced, if the micelles become more extended laterally. As a consequence, the tensions at the ends of the micelles increase up to a certain finite limiting value of the thickness of the micelle characteristic for the fully grown lamella. The said nuclei are assumed to be descendants of local alignments. The stability of those alignments depends on the balance between the influences of their slenderness and of their length. Both properties are distributed in the melt by virtue of statistical rules. The longitudinal growth of a lamella can only set in, if the undercooling is sufficient for the creation of a sufficient negative free energy difference per unit cross-section in the interior of the lamella (with respect to the melt). The value of this negative free energy difference must be larger than that of the positive tensions at the ends of the micelle. This mechanism also explains, why spherulites can melt and leave slender and—nevertheless—stable micelles behind (self-nucleation). In fact, those slender micelles possess a lower positive tension at their ends.

In this way one can also explain, why the number density of nuclei of average lengths increases so tremendously with decreasing crystallization temperature. With iPP and with iPS enormous increases of these numbers could be observed. Factors around one hundred thousand were found, if the melts were quickly undercooled to temperatures 130–160 K below the equilibrium melting points. This has been described in previous sections. Also the influence of pressurization on the number density of effective nuclei can be explained along this route.

Also the nucleation of HDPE can be understood. Enormous number densities of nuclei have regularly been found. The molecules of HDPE are extremely flexible, when compared with other polymers. But this means that the assimilation of the conformations of neighboring molecules is particularly easy.

It should also be emphasized that the large influence of flow on the number density of effective nuclei can be explained by the presence of fringe micelles. Without any doubt these micelles are oriented and their tassels are "ironed" by the action of flow. A closer look on this subject will be preserved for the next section. This comment is only made here for showing that there is enough reason to believe in this concept. Fortunately, the CALTECH group [7] has given a direct proof for the correctness of our assumptions with respect to the nature of the said nuclei. This work has gratefully been quoted in Sect. 2.3.2.3 of this monograph. Probably, the mesomorphic initial states of lamellar growth, as found by Strobl [8] and by Lotz [9], can be harmonized with this view.

Another aspect deserves the attention of the reader. As a starting point for the description of growth speeds the validity of an equation of the type of Eq. (2.14) was proposed in Sect. 2.3.1. At this occasion it has been told to the reader that for primary nucleation, when happening at small degrees of undercooling, such an equation cannot be used. In fact, sporadic primary nucleation for ideal crystals does not take place in polymers, as explained above. Unacceptably long waiting times would be required because spontaneously stretched conformations must be so extremely rare (Flow alleviates this situation). And with lowering the temperature one quickly arrives in a range of temperatures, where kinetic barriers become smaller than kT. But this means that spinodal decomposition, omitting previous associates, can become due.

Some readers of the old school may find this reasoning too digressive. In this respect it is important to have some more formal arguments. As is well-known, Ziabicki [10] proposed an equation, which is based on the classical work by Turnbull and Fisher [11]. This equation reads in the form used e.g. by Chen et al. [1]:

$$\frac{dN}{dt} = C\, kT\, \Delta G\, \exp\left(-\frac{E_a}{kT}\right) \exp\left[-\frac{K_n}{T(\Delta G)^n}\right] \tag{4.1}$$

In this equation dN/dt is the rate of nucleation, which refers to the formation of primary nuclei in the homogeneous melt (exponent $n = 2$) or to the formation of secondary nuclei on the surface of already existing crystallites or on primary nuclei (exponent $n = 1$). The symbol ΔG gives the value of the difference between the bulk free energies of the melt and of the body of the nucleus. For small degrees of undercooling one has:

$$\Delta G = \Delta H\, \frac{\Delta T}{T_m}, \tag{4.2}$$

where ΔH is the latent heat of melting, ΔT is the undercooling and T_m is the equilibrium melting point. In addition one has in Eq. (4.1) with E_a the activation energy and with K_n a constant of rather complicated structure, namely:

$$K_n = \frac{v_0}{nk} \left[\frac{n\, c_{n+1}\, \sigma_{n+1}}{(n+1)\, v_0^{1/2}} \right]^{n+1}. \tag{4.3}$$

In this equation v_0 is the volume of a kinetic element, c is a shape factor and $\sigma_{n+1} = (\sigma_e \sigma_s^n)^{1/(n+1)}$ is the average surface tension, with subscript e standing for the so-called end surfaces and subscript s for the side surfaces. Obviously K_n stands for the energy barrier (the surface tensions included). Except for the fact that in K_n the required parameters cannot be determined readily, there is also the principal objection against the use of this equation for the exponent $n = 2$. In fact, this exponent is reserved for the primary nucleation in the homogeneous melt. In the contrary, Eq. (4.1) can be useful for secondary crystallization, if as the melting point the melting temperature of the spherulites is used. This has been done by Van Krevelen [12], as explained in Sect. 2.3.1.

Unfortunately, Eq. (4.1) is not very useful in the case of the influence of flow. This will be explained in detail in the next section. For the moment it seems sufficient, if it is stated that many experiments on flow induced crystallization have been carried out at temperatures, which are too low for the validity of this equation. And with the aid of flow birefringence etc. it has also been shown in Sect. 3.2 that mostly also the entropy change, as caused during flow by orientation, has been too small for a noticeable increase of ΔG.

We have emphasized from the beginning of our research that the progress of crystallization is proportional to the momentary total surface (the "inner" surface) of the already formed crystalline volume elements. And with isothermal crystallization this growth occurs from the very first moment on the surfaces of nuclei, which are there already. But usually such a momentary inner surface very much depends on the thermal history. With fast cooling many more and smaller nuclei are involved than with slow cooling. But this means that with fast cooling a larger inner surface is created. However, with slow cooling the nuclei, which are activated already at higher temperatures, will get the time to grow over space, where otherwise more smaller nuclei are created later on.

However, there were proposals that the volume fraction itself should be responsible for the progress of crystallization (see e.g. Malkin [13]). These authors argue (correctly) that crystallization is a self accelerating process, which is stopped only by space filling.

If inside the newly formed spherulites secondary crystallization follows with some delay on primary crystallization, there will be a problem with the delayed or incomplete release of latent heat, being responsible for the course of local temperatures. However, for the morphology, as observed on cross-sections, secondary crystallization is of no direct influence. In fact, this morphology is determined by the original number of activated nuclei and by the corresponding number density of spherulites or, in the case of flow, also by structure elements, which are created during this flow. These features do not change any more, if secondary crystallization sets in later on after preliminary solidification.

Notwithstanding all these facts the final degree of secondary crystallization can have a large influence on the properties of the product, in particular on its density and its elasticity modulus. With respect to the fracture mechanics one cannot be so sure of a positive effect. Anyway, under processing conditions the progress of secondary crystallization cannot easily be followed. For instance, a quench can preserve the already existing morphology, but certainly cannot stop secondary crystallization. It can even promote it. And thermal or X-ray methods will be applied at a much too late moment. Curiously enough, a simple method, as proposed by Magill [14] a long time ago, may be helpful in this respect. This author claimed the usefulness of the depolarization effect, which can be determined, when crossed polars are used. It should be indicative for the onset of crystallization. During later investigations on shear induced crystallization also light scattering experiments were used. These investigations will be dealt with in the next section. It turned out that Magill's assumption is not completely true. The depolarization effect may become of importance only during secondary crystallization, as soon as the anisotropy of the scattering units becomes fully developed.

4.2   Views on Flow Induced Processes

The most impressive instigation for own work in this field must be described here: As early as in 1985 the conspicuous relaxation behavior of shear induced crystallization was discovered in Linz, when a sample of PP was investigated (see Sect. 3.3.3.1). But also in our own group the importance of this discovery escaped notice for almost five years. As a consequence the diploma thesis by H. Wippel [19] seemed not very peculiar. For his doctorate a different subject was chosen. Finally, however, this scientific adventure formed the starting point for all of our ideas about flow induced crystallization. The pertinent facts were published twice in original papers [15, 16] and later in two reviews of 1990 and 1997 [17, 18]. But, does it wonder, if other authors ignored our finding completely? So the pertinent author was happy, when Alfonso and his people finally grasped the door-latch around 2000 (see e.g. [20]).

Apparently unstable thread-like precursors, which were formed under high shear loads at high temperatures (not too far below the equilibrium melting point), relaxed in the quieted down melt, when these temperatures were not lowered. But the pertinent relaxation times increased with decreasing temperatures of shearing much faster than the relaxation times of the spectrum of free or entangled molecules. At the much lower melting temperature of the spherulites, the relaxation time of the said threads had increased practically to infinity. As in these experiments duct flow was applied, the shear rate decreased with the distance from the walls. As a consequence a fixed distance from the wall had to be chosen for these experiments. In the course of those experiments the method of short term shearing (or stretching) was invented by us. Sometimes the term "shear pulse" has later been used by others.

Also, it was evident that we had obtained threads. After proper quenches the traces of these threads could be seen by TEM in cross-sections parallel to the previous flow direction. The mutual distances of these traces could better be seen in cross-sections perpendicular to the previous flow direction at the above mentioned special distance from the wall. The number of locations of star-like appearance, where the threads pierced the cross-section, could easily be counted (see Fig. 3.10). Rather big mutual distances could be observed. From these distances one could calculate the total length of the threads per cubic meter. Enormous macroscopic lengths were found. It was evident to us that these threads could not have been formed by lateral association of neighboring oriented macromolecules. Lateral self-diffusion would have taken much too long times, when compared with the short shearing times. In fact, mutual distances of about 1 μm should have been bridged in a transverse direction during the period of flow, a realization, which seemed impossible. The conclusion was that the said threads grew longitudinally during the period of the flow from spots in the melt, where nuclei of some kind had been formed. This has been illustrated by Fig. 3.15. See also the end of Sect. 3.3.4.4.

By the way, any influence of lateral self-diffusion becomes improbable also by the fact that shear can act in a destructive way. This fact becomes evident, if already formed aggregates, which seem ready for association, are too far from each other in a lateral direction. In fact, those aggregates draw away from each other. The well-known process of disentanglement, which causes the non-Newtonian viscosity in polymer melt rheology, is a good example for this destructive action of flow. These facts convince us that theoretical approaches, which take into account only the degree of orientation of singled out macromolecules, cannot lead to a realistic description of the process of flow induced crystallization. To the regret of the present author pertinent papers by the research groups in Eindhoven and in Naples must be dismissed as unrealistic [21–23].

In fact, there are two steps happening. In first instance nearly pointlike associates are formed. In succession theses associates form the cross-links of a network, which can easily be stretched. This process must cause a rather strong orientation of the connecting strands. This order of the events (i.e. nucleation first, stretching and orientation as a subsequence) has been documented quite recently in a publication by Janeschitz-Kriegl et al. [24]. This apparatus was no longer containing a duct of rectangular cross-section of a large aspect ratio. In this apparatus a uniform shear rate was produced. For the purpose a new type of an apparatus of the rotational type was constructed. In preventing detrimental secondary flow (in radial direction) the curvature of the flow lines had to be reduced drastically. The required shear flow was created between two discs with surfaces, which were at a distance of the order of one millimeter from each other. One disc was set in rotation for creating the shear rate. The sample, which had to be inserted in its solid state, had to be cut as a ring from a sheet. This sheet was prepared on a press. The radii of this ring varied from seven to nine centimeters. So its medium diameter was eight centimenters (see the previous chapter).

The authors have found that at a temperature close to the equilibrium melting point of iPP (at about 210 °C), and with a shear rate of only 10 s^{-1}, one nevertheless can get a highly oriented crystalline structure. For the purpose the treatment must only be continued to high enough total shears (at unchanged shear rate). There is no doubt that this type of investigation is of particular importance because of the fact that processing always starts at temperatures close to the thermodynamic melting point, where the nucleation starts. The high total shears, which can be obtained with this machine, are important with respect to the widening of the experimental window. Only this enlarged window can show the actual processes happening. In fact, these processes always start in the high temperature range.

For a series of experiments with iPP a temperature of 190 °C was chosen. Interestingly enough one got a spherulitic structure, if a total shear of only 80 was applied. Of course, the sample had to be quenched afterwards. After higher total shears, however, one got stable highly oriented structures. The required shearing could be continued up to total shears of about 3000. This is a remarkable result, if one realizes that steady state flow of the pure melt is achieved already after about five shear units. Apparently, primary associates are created continuously during the shearing. If the amounts of work, which have been applied recently in the new machine, are compared with those, which have previously been applied at high shear rates and short shearing times, one does not find large differences. It is only that low loads were combined with high deformations recently, whereas high loads were combined with low deformations at earlier times. But one should not forget that both types of experiments were carried out in the high temperature range close to the equilibrium melting point.

At this point the reader must be reminded of work, which has been carried out by Prof. Van Puyvelde and his cooperators at Leuven university in Belgium [25]. These authors investigated the flow induced crystallization of a sample of iPP. For the purpose a shearing cell, which had been developed at the Solvay Central Laboratory, was used. Steps of shearing were applied. In some respect this work can be considered as a forerunner of our recent work. In fact, these authors could observe that only with an increased shear treatment oriented crystallization occurred as in our investigation. But there were two drawbacks. The extent of shearing was limited and this shearing occurred only at 145 °C. As a consequence the interesting prehistory, which occurs at much higher temperatures, did not come into action. Our machine would enable all the omitted investigations. However, there is a limit for the work in Linz. This limit is set by the advanced age of the proponent.

Some years ago one of the authors found the remains of a plastic beaker in the mountains of South Tyrol. The side-wall of this beaker had been corroded by the influences of the environment. Only a palisade of bristles (like very thin tooth picks) had survived. All of these bristles had the same length equal to the previous height of the beaker and were grounded in the rim of the bottom, which still existed. Apparently, these bristles were highly crystalline.

The Genova group was able to elucidate the nature and the relaxation behavior of the thread-like precursors. This work was discussed in Sect. 3.3.3.2. Apparently, the said threads contain crystalline pre-stages, in which the macromolecules are arranged without being troubled by entanglements. It is evident that during the formation of such an arrangement the macromolecules are lined out by the flow.

Arrangements, which are bare of entanglements, are also found in the crystalline structure of HDPE samples, which are directly formed during polymerization. Conclusions were drawn from the typical activation energies. These activation energies, which are a factor ten higher than those characteristic for normal viscous flow, are observed for melting processes, if these processes occur at temperatures of small distances from the equilibrium melting point. In such a process molecules are separated bit for bit from the outer surface of the crystalline aggregates. Very close to the equilibrium melting point the said crystals of HDPE decompose practically at once (with an extremely high activation energy, see Sect. 3.3.3.2). Probably, pre-stages of crystallization have been characterized for iPP also by Li and de Jeu [26] in their X-ray studies.

With respect to our view on the specific work as a useful parameter some additional remarks may be of interest. In Sect. 3.3.1.4 a kinetic interpretation of the action of the specific work was given. In this interpretation the role of the rate of shear was mentioned. In fact, shearing enhances the probability for momentous encounter. But this is not an energetic view. In their calculations Zuidema et al. [27] replace the specific work by the stored free energy (or the equivalent recoverable strain). Unfortunately, the present author cannot agree with this replacement. Admittedly, in a transitional situation shortly after the onset of flow, such a calculation may lead to a realistic result. During this transition specific work is successively accumulated in the melt. However, this stored free energy reaches a limiting value in steady flow (after only five shear units!). But our interpretation is independent of this condition. In fact, it turns out that the creation of precursors for crystalline structures is continued also, when the fluid environment has reached its steady flow behavior. One has only to look on Fig. 3.44. In fact, for this PB-1 of a rather low molar mass steady shear viscosity was reached long before the upturn of viscosity occurred. Even for the also investigated PB-1 of the highest molar mass steady state was reached just before the said upturn occurred. Only with extensional flow the upturns occurred for this polymer already during the transitional situation (see Fig. 3.47). At this occasion an important difference in the behavior of shear flow and of extensional flow can be observed. In fact, with shear flow neighboring macromolecules of a selected macromolecule are removed from that macromolecule almost completely after five shear units. With extensional flow, however, they remain in the vicinity of the selected macromolecule for quite a long time of the deformation. This fact seems to be the reason, why a drastic upwards trend of the viscosity sets in long before steady flow is reached.

Also Elmoumni and Winter [28] reported that "steady shearing conditions were amply reached during pre-shearing" in all experiments on their three PP samples. In fact, the Deborah numbers $De = \tau/t_s$ were always very small (with τ being the relaxation time of the melt and t_s being the shearing time). It is that these authors gave an interpretation to their results, which deserves our sympathy: They claimed that the strain requirements for shear induced crystallization were higher than those for the steady state flow. Pogodina et al. [29] had found also for themselves that shishs relaxed very slowly.

The present author is afraid that the authors of ref. [28] are too much impressed by their Weissenberg number, which is defined as $We = q\tau$. In fact, following them many authors did not spare their efforts in looking for the best value of the relaxation time τ, trying hard to use the arguments of Doi and Edwards [30] and their successors. However, in our opinion any relaxation time of the fluid is inadequate for the just mentioned problem. In fact, here we have to do with the relaxation of the formed structure elements and not with the relaxation of the fluid. And we have learnt that in most cases the relaxation time of the structure elements is by far larger than the relaxation times in the spectrum of the fluid at the temperatures chosen. And if the relaxation time of the structure elements is by far larger than any shearing time t_s, one can add up in principle the effects of prolonged shearing. The total amount of the specific work will be essential. The Weissenberg number can only be a measure for the readiness of association, even if in the mostly non-linear cases the seriously reduced relaxation time of the fluid is not equal to the time needed for the establishment of an oriented structure.

So far it has been assumed frequently that between the applied specific work and the rate of activation of nuclei a correlation exists, which is specific for the type of polymer chosen. The position of the line in Fig. 3.24 represents this correlation for iPP. We were able to show this with the aid of another, high molar mass industrial iPP (see Stadlbauer et al.[31]). In a plot of the number density of nuclei versus the specific work the points for the high molar mass PP coincided readily with those of the general purpose PP.

However, it will be shown immediately that this consideration must be modified, if polymers of unusual molar mass distributions are investigated. With industrial polymers the molar mass distribution has always a similar shape. This means that the experiences gathered so far can be used for many industrial polypropylenes without large modifications. The effect of varying average molar masses is then almost automatically taken into account. However, Kumaraswamy, Kornfield, Yeh and Hsiao [32] used also exotic mixtures of polypropylenes, in which small portions of a fraction of very high molar mass were contained. They also investigated the behavior of the pure high molar mass fraction. They found that in the mixtures the presence of the high molar mass fraction accelerated the growth of shishs considerably. This fact can easily be understood qualitatively. Surprisingly however, the pure fraction of high molar mass did not produce shishs at all at similar shear stresses. According to the well-known behavior of elastic liquids, for which external stress and birefringence are proportional, if the stress is not exceptionally high, there is no doubt that at a certain usual shear stress the average orientation is the same in a sharp fraction of high molar mass and in a polydisperse polymer, when these polymers are characterized by a molar mass distribution containing a high molar mass tail. However, in the sharp fraction orientation will evenly be distributed over all molecules, so that each molecule is only slightly oriented. In the polymer, which contains a high molar mass tail, however, mainly the longest molecules are oriented. They are stretched extraordinarily and contribute a major part to the rheology. So, they also preferentially further the growth of the shishs (according to Fig. 3.15).

This comment, however, does not mean that we can return to a description, in which only the degree of molecular orientation plays a role. The usual calculations make use of an equation of the type of Eq. (4.1). In such an equation the under-cooling is reflected by ΔG. In principle, this ΔG is increased by the decrease of the entropy, as caused by the orientation of the free or entangled molecules. However, in Sect. 3.2 it has been shown that the corresponding increase of ΔG will not be large enough. In fact, it can be measured with the aid of flow birefringence. However, even at the highest attainable shear rates it remains insufficient for a considerable increase of the melting point. Evidently, this judgment is based on the assumption that stresses in polymeric fluids are of a purely entropic nature (no energetic contribution by-say-hindered rotations etc.). But this is a very common feature of elastic liquid models (polymer melts). Naturally, also advanced model calculations adhere to the entropic nature of the stresses. However, these theoretical models are based on additional simplifications (use of dumb-bells or of the reptation model). The measured flow birefringence, however, is free of this ballast.

Admittedly, close to the equilibrium melting point strong shearing actually produces shishs, as has been shown already in our early experiments. Probably, however, one should seek the explanation for this formation of shishs in the fact that the macromolecules are uncoiled, so that they are ready for the necessary association (in a reduction of the activation energy E_a). This is in contrast to the quiescent melts, where one cannot find neighboring macromolecules of stretched conformation. As even with a strong shearing the equilibrium melting point is hardly increased, one is allowed to draw conclusions about the position of this equilibrium melting point. Two examples have been discussed in previous sections, namely for isotactic polypropylene and for isotactic polystyrene. In fact, for both polymers the relaxation time of the shishs approached the value zero, when the known equilibrium melting point was approached. For the stable Form I of i-PB-1 this equilibrium melting point was never determined to our knowledge. However, as we have concluded that the relaxation time of the shishs approaches zero near 170 °C [33], one can assume that in the vicinity of this rather high temperature the equilibrium melting point of Form I of this polymer will lie. The melting temperature of the spherulites of Form I lies at about 130 °C. So one finds the usual undercooling of about 40 °C, as depicted in Fig. 1.3 for iPP. Unfortunately, a detailed determination of the crystallization kinetics under the influence of flow at the high temperatures near the equilibrium melting point has not yet been possible for this polymer.

It should be emphasized that in the experiments of other authors, which were reviewed in Eder, Janeschitz-Kriegl and Liedauer in 1990 [16], only mild conditions of shearing were applied. In these investigations the samples were sheared up to the moment, when signs of crystallization were obtained (viscosity upturn, light scattering etc.). Also, part of the work by Hadinata et al. [34] was carried out at those mild conditions. Compared with these conditions a series of conditions applied in Linz [17] must be considered as very harsh. In particular, shishs occurred only at comparatively high mechanical loads (see e.g. Janeschitz-Kriegl, Ratajski and Stadlbauer [35]). Apparently, these puzzles can be solved with the aid of the recently developed apparatus [24].

So far the influence of nucleation agents has hardly been mentioned in the present monograph. At the time Binsbergen [36] was seriously engaged in this matter. In particular, if a sorbitol derivative is used as an additive in iPP, one finds quite intriguing results, as also very recent work by Balzano [37] shows. (At Linz we are very much indebted for quickly receiving this doctoral thesis from Eindhoven university). The interaction of a compound like bis(dimethyldibenzyli-dene)sor-bitol (BDMDBS) with iPP has previously been discussed by Thierry et al. [38], Alcazar et al. [39] and Kristiansen et al. [40]. BDMDBS is used as a clarifier in PP, which means that it normally produces an enormous amount of tiny spherulites with diameters below the wave length of the light. The central part of the flat molecules of BDMDBS is polar, whereas the fringes are apolar. This compound dissolves in the melt of iPP above 250 °C. Concentrations lower than one percent are usual. With cooling BDMDBS crystallizes from such a solution at a tempera-ture, which depends on the concentration, but always lies above the equilibrium melting point of iPP (See also [41]). With this phase separation fibrils are formed, which become part of a network. In the fibrils the molecules of BDMDBS are concentrated. The necessary cohesion is caused by hydrogen bonds between the central parts. The outside of the fibrils is apolar. The distance between the molecular layers corresponds with the 3_1 helix of iPP. With short term flow the network is destroyed. At the same time, the fibrils should be oriented and act like thread-like precursors for lateral growth of lamellae of iPP. As always, this type of growth happens after cessation of flow. In the mentioned thesis by Balzano many details are explored, i.e. by small angle X-ray Synchrotron scattering, by dynamic mechanical measurements and by differential scanning calorimetry.

Nevertheless, some question marks will be unavoidable. In fact, only one selected shearing condition was applied, namely a standard short term shearing for 3 s at a shear rate of 60 s^{-1}. This choice is certainly insufficient. Our experience is that in pure iPP the mechanical load is one of the most important parameters. In Region I, at temperatures above the phase separation of BDMDBS, only isotropic structures were found after the quench to room temperature. However, at suitable thermal and mechanical conditions similar effects can be found also in the absence of BDMDBS. In this connection one should look at our Figs. 1.1, 3.23, 3.24. In all these cases seemingly point-like nuclei were caused by the influence of flow. In reality, however, these nuclei were not point-like. They only had an average length smaller than their mutual distances. As a consequence, spherulites could grow on them after cessation of the flow. As a consequence isotropic morphologies were created. An important additional information is that the number density of these activated nuclei is growing tremendously with increasing shearing load (See also Tribout et al. [3]). Such an investigation is missing in the thesis by Balzano. Probably, one will find a similar effect in the presence of BDMDBS. This effect can even be more pronounced with BDMDBS. It can also give disclosure on the stability of the fibrils in the flow field. In fact, it may be that these fibrils are broken in stronger flows because of the relative weakness of the hydrogen bonds and their comparative thickness, as compared with normal polymer molecules. Because of this thickness these fibrils will be particularly prone to the forces exerted by the flow. In fact, these fibrils possess a relatively large surface per unit of length. By the fracture of fibrils the number density of effective nuclei can be increased enormously. Unfortunately, a count of the number of nuclei or of corresponding spherulites has not been reported. But it seems that this criticism is not justified because of the fact that the load of work becomes unbearable.

In pure polymer melts, the formation of thread-like precursors, which are built of polymer molecules only, is furthered by increasing shearing loads (see the two open triangles in the upper right corner of Fig. 3.23). Those thread-like precursors can grow even close to the equilibrium melting point of i-PP. They decompose quickly at these high temperatures after cessation of flow. Only a fast quench can preserve them (see Sect. 3.3.3 and Fig. 2.20). In fact, according to Balzano [37] also the influence of BDMDBS vanishes in all his experiments, if they are carried out close to the equilibrium melting point of i-PP near 212 °C [42].

As and Maxwell Haas [43] reported already a long time ago, no textural details could be seen under the microscope in their experiments, when the upturn in the viscosity was surpassed. Apparently, the spherulitic morphology, which was found by Hadinata et al. [34] for a similar PB-1 just before the viscosity upturn, was smoothed out by continued shearing (apparently at increased stress). This fact may serve as an excuse for those theoreticians, who still use the Nakamura approach [44], which has been criticized in Sect. 1.2.1 for being inadequate for the prediction of morphologies.

In this respect, the simulation of melt spinning must be mentioned. Obviously, melt spinning is a very complicated process. All kinds of influences have to be taken into account, as there are: the influences of: rapid stretching, heat transfer (stabilizing the process by cooling), lateral shrinkage, air drag, inertia, gravity. According to Spruiell and White [45], very often there is an abrupt uphold in the decreasing fiber temperature at the distance from the spinneret, where crystallization starts. This effect shows that crystallization occurs quite rapidly. Probably it does not matter too much for the interpretation of such a fast process, if the details of crystallization are ignored. With so-called high-temperature polymers amorphous regions can be retained in a crude determination, as will be discussed below. In this connection the present author has to admit that it cannot be his task to review those intricate papers, and if it is only because of his lacking experience with numerical simulations. As a consequence only the last paper of the school of McHugh [46] is quoted. Previous papers of this group can be found in the quoted paper. A similar comment also holds for a recent paper by Tanner [47], even if this author claims that he can describe a slow isothermal process reported by Wassner and Maier [48], for which more details have been made available by Hadinata et al. [34]. The applied Nakamura process, however, is certainly unable to describe details.

A special subject deals with polymers, which crystallize very slowly. An explanation for such a slow process has been given by Van Krevelen (see Fig. 2.18 of the present monograph). In those polymers the distance between the glass transition temperature and the melting point is rather small. Mostly one has to do with so-called high temperature polymers. As a consequence one finds with cooling amorphous glassy areas in the samples. Numerical simulations on the basis of Nakamura's equation can furnish boundaries between amorphous and crystalline areas. For this purpose Nakamura's theory seems realistic enough. The only necessary assumption is that the glass transition temperature, as known a priori, is much less influenced by the cooling speed and by the rate of deformation than the crystallization kinetics. But this means that nobody knows the position of the said boundary exactly. Fortunately very often, the exact position of such a boundary is not so relevant.

However, there is still another feature of these special polymers. Mostly the chain lengths of their molecules are rather small compared with those of—say—the polyolefins. Amongst others this means that the growth speed of the spherulites is no longer independent of the molar mass, as has been shown by Van Krevelen [12] (See also Sect. 2.2.2.2). As a consequence the "demarcation" (see Fig. 1.3), which is so characteristic for long chain polymers, does no longer exist. In fact, this demarcation is based on the fact that in a quiescent melt sufficiently long molecules will never reorganize spontaneously into a stretched conformation. But such a conformation is much more probable for the mentioned short and mostly stiff molecules. So, one must conclude that the crystallization kinetics of the short chain polymers differs in many respects considerably from those of the polyolefins. It looks much more like that of ordinary low molar mass materials. Only the slowness of the process reminds us of the fact that one has still to do with macromolecules. It may be instructive to mention that in the corresponding temperature ranges the growth speeds of the spherulites of PET are about two decades lower than those of PB-1 (see Fig. 2.15), and also that a comparison of the characteristics of flow induced crystallization has not yet been carried out.

Returning to the problem of crystallization in the spin line one has to state that also the cross-sections through melt spun fibers do show details of structure. This has been shown by White and Cakmak [49] in a monumental study quite a long time ago. For the moment, however, the question remains still unanswered, whether those structures can be understood completely. The reports of Linz university have features of a detective story, whereas the just mentioned paper confronts us with solid facts, as obtained in a complex process. We always tried to find simple situations, which are interpretable, but still realistic enough for an industrial process. So the character of research differs very much, and the present author does not feel competent to review papers like the one cited above. But this does not mean that an extended synopsis of those results cannot lead to important insights. Meanwhile Cakmak has published a great number of papers, in which structures are shown in fibers and in injection molded parts. These results were mostly obtained with high temperature polymers, as characterized above. Obviously, the activities of Cakmak were strongly dictated by problems, which came up during the industrial processing of those special polymers or their mixtures. Nowadays it is easy to find papers of M. Cakmak, and also-say-of Avraam I. Isayev-in the internet. And if the present author does not feel prepared for a review, it may be sufficient, if he quotes the names in reminding the reader of the merits of these authors.

214 4 Closing Remarks

At the end of this last chapter I feel the necessity to mention two contributions of the school in Sophia Antipolis. In the first contribution by Haudin et al. [5] an advance has been made to determine the growth speed as a function of the shear rate. As the reader may remember, this subject has not been treated extensively in this book. There it has been concluded that this growth speed cannot be very dependent on the deformation rate. This fact was attributed to the destructive action of flow, which should be particularly pronounced in shear flow. From this point the contribution by Haudin et al. deserves particular interest. These authors have observed that the effective thickness of a glass fiber increases continuously, if it is drawn with constant speed through an undercooled crystallizing polymer melt. Obviously, this growth must have to do with the increasing thickness of a crystalline layer on the surface of the glass fiber under the influence of flow. This should be a particular simple experiment.

From Eq. (3.19) we learn that the shear rate in the fluid must decrease rapidly with the distance r from the heart of the glass fiber. The radius r_f of the fiber is now replaced by the growing apparent radius of the fiber, which for simplicity is now called $r_t(t)$. At various fixed distances r from the heart of the fiber (outside $r_t(t)$) one obtains now an estimate of the flow rates in the fluid, as it varies with time t. Steady state conditions, as reflected by Eq. (3.19), are no longer guaranteed! Anyway, if one observes a point at a fixed distance r in the fluid, one finds a continuous increase of the shear rate up to the moment, when this point is reached by the growth front. But this encounter causes the end of this development. In contrast, this final shear rate, as reached at the arrival of the growth front, decreases with the distance r. If now the observed rate of the apparent radius of the fiber at the growth front would be seen as the required growth rate of the crystalline zone as a function of the local shear rate, one should expect a decreasing function with increasing $r_t(t)$. But this has not been found. Instead, a practically constant apparent growth rate was found.

Correctly, the authors concluded that the fluid was pre-treated, before the growth front arrived. One reads: "Crystallization occurs under a complex shear rate and shear stress history and nevertheless the growth rate is constant." In fact, in the beginning of the experiment, when the apparent radius $r_t(t)$ is still small, the polymer melt passes through high shear rates acting only for a short time span. Later, when the radius $r_t(t)$ has become relatively large, one has much lower shear rates, but acting over a much longer time.

As the reader may remember, these ideas are not unfamiliar to us. But these experiments show that it is almost impossible to get reliable values of the growth rate as a function of the local shear rate. In this respect we cannot agree with the second part of the above sentence, as quoted from the original paper. But we can be thankful to the authors for demonstrating their results so clearly. With a large number of polypropylene samples of widely varying parameters they found for the sensitivity to shear the following equation:

$$S = 0.65 \ \exp(7.7 \times 10^6 \ M_w), \tag{4.4}$$

where S is the ratio of the growth speed under shear over the growth speed in the quiescent melt. This equation holds for a temperature of 126 °C and does not contain the shear rate as a parameter. For us the role of the molar mass M_w remains unclear. We know that the number density of nuclei always increases rapidly even with low shear rates (with low specific mechanical works). But the rate of this increase depends on the viscosity, which—on its part—is a function of the molar mass. If those nuclei are already in the fluid, when the growth front at the fiber arrives, one must not be surprised, if that growth rate is influenced by M_w.

Finally, however, there is a message, which will be much more in favor of the group at Sophia Antipolis. Together with a group at the Katholieke Universiteit Leuven an investigation was carried out with the aid of optical means (Devaux et al. [50]). In this context the use of small angle light scattering (SALS) is of interest, in particular for isotropic samples. According to the theory by Stein and Wilson [51] one has to measure the total scattered intensity I_{Vv}, which is obtained, when parallel polars are used, and the total scattering intensity I_{Hv}, when crossed polars are used. Scattering invariants were defined, which are given by:

$$Q_\eta = \int_0^\infty \left(I_{Vv} - \frac{4}{3} I_{Hv} \right) q^2 \, dq \tag{4.5}$$

and

$$Q_\delta = \int_0^\infty I_{Hv} \, q^2 \, dq. \tag{4.6}$$

In these equations q is the scattering vector defined as:

$$|q| = \frac{4\pi}{\lambda} \sin \frac{\theta}{2}, \tag{4.7}$$

where λ is the wave length of the light and θ is the polar scattering angle. Invariant Q_η is proportional to the mean-square density fluctuation $<\eta^2>$ and invariant Q_δ is proportional to the mean of the squared anisotropy $<\delta^2>$.

Measurements with home made optics, as combined with the Linkam unit (see Mackley et al. [52] and also Sect. 3.3.5.1), furnished a clear picture of the development of crystallization in the undercooled quiescent melt of an industrial PP. This picture is shown in Fig. 4.1.

The most important result of this picture seems to be that the invariant Q_η for the density fluctuations reaches already its maximum, when the invariant Q_δ for the internal anisotropies just starts to grow. Without any doubt the course of this latter invariant indicates the ripening of the structure. On the other hand, at the time,

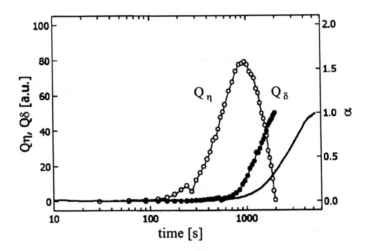

Fig. 4.1 Evolution of the scattering invariants Q_η and Q_δ in a quiescent melt of an i-PP at a temperature of 136°C according to [51]. The full line gives the course of the degree of depolarization. Courtesy of Springer Verlag

when the maximum of Q_η is reached, about half of the volume is already occupied by the embryonic phase. And the true depolarization starts even later than Q_δ (notice the logarithmic time scale!).

Certainly, we may assume that the upcoming of strong anisotropies have to do with the secondary crystallization. This has already been emphasized at the end of Sect. 4.1. If a series of shear flows of increasing shear rates is applied, the time scale shrinks tremendously. The authors found that with an unchanged shearing time of 30 s and increasing shear rates up to 1.4 s^{-1} the location of the maximum in Q_η shifted from about 2000 s at zero shear rate to about 400 s at 1.4 s^{-1}. But a shear rate of 1.4 s^{-1} must still be considered as a very low shear rate. However, the present author (Janeschitz-Kriegl [53]) has been able to show that an interpolation leads to results, which fit quite well with those obtained in Linz at much higher shear rates.

Figure 4.1 is explicitly shown here as a hint for future research, even if one must not forget in this connection that Eqs. (4.5) and (4.6) only hold for the isotropic case. For the present presentation, however, this does not matter. In fact, the crystallization of Ref. [50] occurred after the cessation of a flow, which was mild enough for the occurrence of spherulites. If one would aim at an investigation of the course of secondary crystallization, as occurring already during flow, one should start with a systematic investigation of light scattering and depolarization. For the purpose these measurements should be carried out at a series of temperatures, pressures and deformation rates for all kinds of interesting polymers. Because of the application of deformation rates calorimetric measurements will not be feasible. With respect to the optics one will have to be content with simpler concepts.

The course of secondary crystallization is of interest in connection with the release of latent heat in processes, in which heat transfer is involved. In particular, the course of this release can be important, if numerical simulations are to be carried out. So far numerical simulations of—say—injection molding were carried out without this knowledge. But inadequately calculated local temperatures can give a wrong picture. Sometimes one can be lucky, if secondary crystallization sets in only after the form giving process is finished. But in such a case the puzzle remains about the part of the latent heat, which is relevant already to the primary ("embryonic") crystallization. However, the present book should not have been written, if this primary crystallization would not be so relevant for the formation of decisive structures.

Unfortunately, crystallization kinetics in equi-biaxial extension has not yet been investigated. It can be very different from the kinetics in uni-axial extension.

4.3 Final Résumé

On the pages of this book quite a number of access routes have been described. Nevertheless, one has to admit that a universal route for the description of the solidification of polymers by crystallization, as happens during processing, cannot yet be given. In fact, the behavior of these polymers can be rather complex. A practicable way for the elucidation of pertinent puzzles seems to be description of the situations coming up in the three processing techniques of overwhelming importance, as there are injection molding, extrusion and melt spinning.

Let us start with the first subject. There is no doubt that there are at least five parameters. One has in turn: The temperature of the polymer melt, before it enters the mold, the speed of the injection, the temperature of the walls of the mold, the final pressure increase guaranteeing the desired shape of the sample and the cyclus time, the latter being to some extent a function of the other parameters. But a proper choice of these parameters can considerably differ from polymer to polymer. In this connection the mentioned choice can be facilitated by a comparison of the crystallization speeds and temperatures of relevant polymers. Interestingly enough the crystallization speeds of three polymers possessing a carbon-carbon backbone, as there are linear polyethylene, isotatic polypropylene and isotactic polystyrene, differ tremendously. The first mentioned one has a speed of crystallization, which is a factor thousand higher than that of the second one. But a similar factor holds between the second and the third polymer. The reader should tell the author a technology, which bridges these factors. But there is still another complication. The mentioned factors were determined for the quiescent melts. But flow, which is essential in polymer processing, can change these factors tremendously. Increases of the crystallization speeds, as caused by the flow, by factors of several hundreds must be taken into account. However, there do not exist enough reliable data. Also flow and the final increase of the pressure will influence the morphology of the sample.

But there is also another point of consideration, which has to do with heat transfer. Practice requires a temperature for the polymer at the entrance to the mold, which is much higher than the equilibrium melting point of the polymer. Otherwise one can get congestions at narrow passages in the mold, where an increased flow rate can cause rash crystallization. This danger has to do with the fact that the temperature of the mold must be well below the melting point of the polymer. Fortunately, however, solidification by crystallization is not an immediate process. There is always a delay, as described in this report. But this delay becomes shorter with a lowering of the temperature of the mold wall.

However, there is still another point of consideration. It appears that during the short period of flow usually only a restricted number of thread-like precursors got the chance to grow. It can also be that some of the starting points are formed by dust particles, which cannot be avoided completely. As has been shown in our investigation, secondary growth in a direction perpendicular to the said threads, happens preferentially only after the cessation of the flow. But an increasing rate of secondary growth has been found mainly with an increased cooling rate.

Some experimental approaches must also be discussed. So far it appeared that a controlled continuous decrease of temperature could not be realized. As a consequence researchers tried to work at several fixed degrees of supercoolings, which were chosen rather close to each other. Unfortunately, with those experiments one can get with increasing supercoolings wrong orders of steps, namely first cooling and only afterwards shearing. However, such a sequence will never occur in practice. Nevertheless one will learn from those experiments. But one must keep in mind that also rare things can happen, in particular, when a duct of a rectangular cross-section of a large aspect ratio is used. One example is the occurrence of distinct highly oriented layers close to the duct walls, when inside the duct spherical structures are obtained. Such an effect will never be found after a continuous cooling of the still streaming melt, as happens during filling.

Another effect has been found after extended shearing between parallel metal plates, when one of these plates was kept stationary and the other plate was moved in creating a homogeneous shear rate. Admittedly, however, such a process cannot be realized unrestrictedly, if the mobile wall can be moved only in a strictly straight direction. In such a case the totally realizable shear is very restricted because of the finite lengths of the metal strips. But one can escape this situation by the application of an approximation. For this purpose a certain not too large curvature of the flow lines has been permitted. In such a concept secondary flow, which becomes of importance with an increased degree of shearing on curved flow lines, is minimized. In usual rotational viscometers the said curvature appears to be always too strong. Admittedly, however, a prolonged shearing, as is required for the development of flow induced crystallization at sufficiently high temperatures (close to the melting point), does never occur in usual rheometry. But the application of such an increased degree of shearing widens the experimental window considerably, even if a practical application is not intended. In fact, a better insight into the occurring processes can be obtained along this route.

If with the use of a screw extruder the processes happening in and after the die are considered, no mold walls can be removed after solidification. In fact one has a more complicated situation. The separation from the wall of the die occurs, when crystallization has at least partly occurred.

There is still another point of interest. It has been shown by us that during a shear flow with a reasonably large shear rate steady flow is reached almost immediately, i.e. after about five shear units. As crystallization processes do not occur immediately, this statement also holds for crystallizing polymers. With extensional flow, however, the situation seems quite different. Whereas exchange of neighboring macromolecules happens very fast in shear flow, the situation with extensional flow is rather different. In fact, with extensional flow neighboring macromolecules remain close to each other for a longer time span. As a consequence crystallization can start, before steady flow is reached (see Fig. 3.49). Unfortunately, for biaxial extension (see blow molding) no experiments with respect to crystallization seem to exist.

This short review is given in order to encourage researchers to continue similar research activities. Unfortunately, general interest has shifted away from polymer science to other subjects. But in this connection it must be emphasized that the role of polymers is still evident. Nevertheless the author had to feel the change, when he had to stop his activities officially, when he reached the age of seventy years about two decades ago. Fortunately he could continue to some extent with the aid of the Austrian Science Foundation. In this connection the apparatus for extended shearing was built in the Netherlands by Messrs. Mouthaan in Delft. Unfortunately, an apparatus for crystallization in extensional flow could not be completed because of a series of difficulties emerging.

With the aid of the said esteemed foundation he was enabled to continue the employment of Dr. Ewa Ratajski for eight further years. Her cooperation was essential for the success. He had also the occasion to discuss problems with Prof. Gerhard Eder, who also contributed some mathematical derivations.

Finally it should probably be mentioned that the gathered experiences are the consequence of about forty years of research. One cannot expect that a young person can gather this amount of experiences within a limited number of years. Also, a restricted view on industrial interests can be useful. But recently industry shows interest only in short term programs. This is in contrast to previous usage, when graduates, after leaving the university, were placed in industry-owned research laboratories, before they were selected for special tasks.

References

1. Chen Q, Fan Y, Zheng Q (2006) Rheological scaling and modeling of shear-enhanced crystallization rate of polypropylene. Rheol Acta 46:305–316
2. Al-Hussein M, Strobl G (2002) The melting line, the crystallization line and the equilibrium melting temperature of isotactic polystyrene. Macromolecules 35:1672–1676

3. Tribout C, Monasse B, Haudin JM (1996) Experimental study of shear-induced crystallization of an impact polypropylene copolymer. Colloid Polym Sci 274:197–208
4. Magill JH (1967) Crystallization of poly (tetra-p-silphenylene) siloxane J Polym Sci A-2, 5:89–99
5. Haudin JM, Duplay C, Monasse B, Costa JL (2002) Shear induced crystallization of polypropylene. Growth enhancement and rheology in crystallization range. Macromol Symp 185:119–133
6. Becker R, Döring W (1935, in German) Kinetic treatment of the formation of nuclei in over-saturated steam. Ann Phys 5(24):719–752
7. Olsen AP, Flagan RC, Kornfield JA (2006) Manipulation of athermal nuclei in aqueous poly (ethylene oxide) by scanning activity gravimetric analysis. Macromolecules
8. Strobl G (2000) From the melt via mesomorphic and granular layers to lamellar crystallites: a major route followed in polymer crystallization? Eur Phys J E 3:165–183
9. Lotz B (2000) What can polymer crystal structure tell about polymer crystallization processes? Eur Phys J E 3:185–194
10. Ziabicki A (1996) Crystallization of polymers in variable external conditions. 1. General equations. Colloid Polym Sci 274:209–217
11. Turnbull D, Fisher JC (1949) Rate of nucleation in condensed systems. J Chem Phys 17:71–73
12. Van Krevelen DW (1990) Properties of polymers, 3rd edn. Elsevier, Netherlands, pp 594–603
13. Malkin AYa, Beghishev VP, Keapin IA, Andreyanova ZS (1984) General treatment of polymer crystallization kinetics—Part 2: the kinetics of nonisothermal crystallization. Polym Eng Sci 24:1402–1408
14. Magill JH (1962) A new technique for following rapid rates of crystallization, II isotactic polypropylene. Polymer 3:35–42
15. Janeschitz-Kriegl H, Wimberger-Friedl R, Krobath G, Liedauer S (1987, in German) On the formation of layer structures in plastic parts. Kautschuk + Gummi, Kunststoffe 40:301–307
16. Eder G, Janeschitz-Kriegl H, Krobath G (1989) Shear induced crystallization, a relaxation phenomenon in polymer melts. Progr Polym Sci 80:1–7. Janeschitz-Kriegl H, Eder G (2007) Same title: A recollection. J Macromol Sci Part B 46:1–11
17. Eder G, Janeschitz-Kriegl H, Liedauer S (1990) Crystallization processes in quiescent and moving polymer melts under heat transfer conditions. Progr Polym Sci 15:629–714
18. Eder G, Janeschitz-Kriegl H (1997) Processing of polymers 5: crystallization. Mat Sci Techn 18:269–342
19. Wippel H (1989) Further investigations concerning shear induced crystallization of isotactic polypropylene, diploma thesis, Linz University, see also Fig. 2.20 (in German)
20. Azzurri F, Alfonso GC (2005) Lifetime of shear-induced crystal nucleation precursors. Macromolecules 38:1723–1728
21. Peters GWM, Swartjes FHM, Meijer HEH (2002) A recoverable strain-based model for flow-induced crystallization. Macromol Symp 185:277–292
22. Van Meerveld J, Peters GWM, Hütter M (2004) Towards a rheological classification of flow induced crystallization experiments of polymer melts. Rheol Acta 44:119–134
23. Coppola S, Balzano L, Gioffredi E, Maffettone PL, Grizzuti N (2004) Effects of the degree of undercooling on flow induced crystallization in polymer melts. Polymer 45:3249–3256
24. Janeschitz-Kriegl H, Ratajski E, Eder G (2014) Unlimited shear as a source of information in polymer melt processing. Int Polym Proc 29:402–411
25. D'Haese M, Van Puyvelde P, Langouche F (2010) Effect of particles on the flow-induced crystallization of polypropylene at processing speeds. Macromolecules 43:2933–2941
26. Li L, de Jeu WH (2003) Shear-induced ordering as a precursor of crystallization in isotactic polypropylene. Macromolecules 36:4862–4867
27. Zuidema HG, Peters GWM, Meijer HEH (2001) Development and validation of recoverable strain based model for flow-induced crystallization of polymers. Macromol Theory Simul 10:447–460

28. Elmoumni A, Winter HH (2006) Large strain requirements for shear-induced crystallization of isotactic polypropylene. Rheol Acta 45:793–801
29. Pogodina NV, Lavrenko VP, Srinivas S, Winter HH (2004) Rheology and structure of isotactic polypropylene near the gel point: quiescent and shear-induced crystallization. Polymer 42:9031–9043
30. Doi M, Edwards SF (1986) The theory of polymer dynamics. Clarendon Press, Oxford
31. Stadlbauer M, Janeschitz-Kriegl H, Eder G, Ratajski E (2004) New extensional rheometer for creep flow at high tensile stress. Part II. Flow induced nucleation for the crystallization of PP. J Rheol 48:631–639
32. Kumaraswamy G, Kornfield JA, Yeh F, Hsiao B (2002) Shear-enhanced crystallization in isotactic polypropylene. 3. Evidence for a kinetic pathway of nucleation. Macromolecules 35:1762–1769
33. Braun J, Wippel H, Eder G, Janeschitz-Kriegl H (2003) Industrial solidification processes in polybutene-1. Part II—Influence of shear flow. Polym Eng Sci 43:188–203
34. Hadinata C, Gabriel C, Ruellmann M, Laun HM (2005) Comparison of shear-induced crystallization behavior of PB-1 samples with different molecular weight distribution. J Rheol 49:327–349
35. Janeschitz-Kriegl H, Ratajski E, Stadlbauer M (2003) Flow as an effective promotor of nucleation in polymer melts: a quantitative evaluation. Rheol Acta 42:355–364
36. Binsbergen FL (1970) Heterogeneous nucleation in the crystallization of polyolefins: Part I chemical and physical nature of nucleation agents. Polymer 11:253–267
37. Balzano L (2008) Flow Induced Crystallization of Polyolefins. Doctoral Thesis Eindhoven University of Technology
38. Thierry A, Fillon B, Straupé C, Lotz B, Wittmann JC (1992) Polymer nucleation agents: efficiency scale and impact of physical gelation. Progr Colloid Polym Sci 87:28–31
39. Alcazar D, Ruan J, Thierry A, Lotz B (2006) Structural matching between the polymeric nucleating agent isotactic poly(vinylcyclohexane) and isotactic polypropylene. Macromolecules 39:2832–2840
40. Kristiansen M, Werner M, Tervoort T, Smith P, Blomenhofer M, Schmidt HW (2003) The binary system isotactic polypropylene/bis (3, 4-dimethyl benzylidene) sorbitol: phase behavior, nucleation and optical properties. Macromolecules 36:5150–5156
41. Jerschow P, Janeschitz-Kriegl H (1997) The role of long molecules and nucleation agents in shear induced crystallization of isotactic polypropylene. Intern Polym Proc 12:72–77
42. Marand H, Xu J, Srinivas S (1998) Determination of the equilibrium melting temperature of polymer crystals: linear and nonlinear Hoffman-weeks extrapolation. Macromolecules 31:8219–8229
43. Haas TW, Maxwell B (1969) Effects of shear stress on the crystallization of linear polyethylene and polybutene-1. Polym Eng Sci 9:225–241
44. Nakamura K, Watanabe T, Katayama K, Amano T (1972) Some aspects of non-isothermal crystallization of polymers. I relationship between crystallization temperature, crystallinity and cooling conditions. J Appl Polym Sci 16:1077–1091
45. Spruiell JE, White JL (1975) Structure development during the melt spinning of fibers. Appl Polym Symp 27:121–157
46. Kohler WH, McHugh AJ (2007) 2D modeling of high-speed fiber spinning with flow-enhanced crystallization. J Rheol 51:721–733
47. Tanner RI (2003) On the flow of crystallizing polymers I. Linear regime. J Non-Newton Fluid Mech 112:251–268
48. Wassner E, Maier RD (2000) Shear-induced crystallization of polypropylene melts. In: Binding DM, et al (eds) Proceedings of the XIII International Congress on Rheology, Cambridge, pp 183–185
49. White JL, Cakmak M (1986) Orientation development and crystallization in melt spinning of fibers. Adv Polym Technol 6:295–338
50. Devaux N, Monasse B, Haudin JM, Moldenaers P, Vermant J (2004) Rheological study of flow enhanced crystallization in isotactic polypropylene. Rheol Acta 43:210–222

51. Stein RS, Wilson PR (1962) Scattering of light by polymer films possessing correlated orientation functions. J Appl Phys 33:1914–1922
52. Mackley MR, Wannaborworn S, Gao P, Zhan F (1999) The optical microscopy of sheared liquids using a newly developed optical stage. J Microsc Anal 69:25–27
53. Janeschitz-Kriegl H (2006) Phases of flow-induced crystallization of iPP: how remote pieces of the puzzle appear to fit. Macromolecules 39:4448–4454

Author Index

© Springer International Publishing AG 2018
H. Janeschitz-Kriegl, *Crystallization Modalities in Polymer Melt Processing*,
https://doi.org/10.1007/978-3-319-77317-9

Subject Index

© Springer International Publishing AG 2018
H. Janeschitz-Kriegl, *Crystallization Modalities in Polymer Melt Processing*,
https://doi.org/10.1007/978-3-319-77317-9

Printed in the United States
By Bookmasters